OUR LIVING WORLD OF NATURE

The
Life
of the
Marsh

THE NORTH AMERICAN WETLANDS

Developed jointly with The World Book Encyclopedia

Produced with the cooperation of
The United States Department of the Interior

The Life of the Marsh

THE NORTH AMERICAN WETLANDS

WILLIAM A. NIERING

Published in cooperation with
The World Book Encyclopedia

McGraw-Hill Book Company
NEW YORK   TORONTO   LONDON

WILLIAM A. NIERING *is Director of the Connecticut Arboretum and Professor of Botany at Connecticut College. It was in the Pocono Mountains of Pennsylvania that he early developed an interest in ecology and subsequently stimulated local interest in the preservation of the Cranberry Bog Preserve. He was educated at Pennsylvania State University and at Rutgers, The State University of New Jersey, where he received his Ph.D. degree in 1952. The author participated as land ecologist in an expedition to Kapingamarangi Atoll in the South Pacific in 1954, and in 1958 he served as conservation consultant to the Regional Plan Association in New York City. The result of his studies in the Greater Metropolitan Area was the publication of* Nature in the Metropolis. *His recent research has involved studies in the mountains and deserts of Arizona and California. In addition to teaching at Connecticut College, he has been associated with the Wesleyan University Graduate Summer School for Teachers. Dr. Niering supervises the Connecticut Arboretum's long-range ecological studies involving the dynamics of terrestrial and wetland habitats. He is active in conservation, especially the preservation of natural areas, and is currently collaborating on a series of interpretive guides to the natural areas of Connecticut for the State Geological and Natural History Survey. The author is a member of numerous professional societies and has published widely in scientific journals.*

Library of Congress Catalog Card Number: 66—28448

567890 NR 7210

46006

OUR LIVING WORLD OF NATURE

Contents

WAYS OF WETLAND LIFE

WETLANDS OR WASTELANDS?

APPENDIX

Our Dynamic Wetlands

All lakes are doomed to die. At first this statement may seem to be an exaggeration, but you may observe this mortality for yourself by visiting any lake or pond. Reach into the water close to the shore and pick up a handful of sediment from the bottom. This black ooze is composed of silt and partly decayed plants and animals that once lived in the water and have gradually piled up on the bottom. As this accumulation of dead material becomes deeper, and consequently the water above it shallower, the lake or pond may turn into a bog, swamp, or marsh. Of course the change takes place so slowly, over many years, that it may seem that nothing at all is happening. But this process of filling in is the chief way most wetlands are formed, even on the coasts, where a similar change produces magnificent salt marshes.

What is so exciting, you may ask, about a swamp or a marsh? The forest and the seashore are pleasant—vacation spots, perhaps, or at least relatively comfortable places to explore the fascinating world of plants and animals. But everyone knows that swamps and marshes can be uninviting indeed, for you may have to contend with swarms of biting

THE DISTRIBUTION OF WETLANDS

The most abundant wetlands, the marshes, are the most widely and evenly distributed. Bogs are concentrated in the colder northeastern and Great Lakes areas. The dots elsewhere on the map indicate the presence of boglike plants rather than bogs. Swamps are lopsidedly concentrated east of the Mississippi River; the more than one-half million acres of mangrove swamps are found only along the coast of southern Florida.

The *inland shallow fresh-water marshes* contain mostly grasses, bulrushes, spike rushes, cattails, arrowhead, pickerelweed, and smartweed. In the North, reeds, whitetop, wild rice, cut-grass, and giant bur reed are characteristic; in the Southeast, maiden cane, saw grass, arrowhead, and pickerelweed. The *inland deep fresh-water marshes*, the best waterfowl breeding habitat, support cattails, reeds, bulrushes, spike rushes, and wild rice; in the Southeast, water hyacinth and water primrose sometimes form surface mats. In the *coastal shallow fresh-water marshes*, grasses, sedges, cattails, arrowhead, smartweed, and arrow arum are the major plants. The *coastal deep fresh-water marshes*, found mainly on the Atlantic and Gulf coasts, have cattails, wild rice, pickerelweed, giant cut-grass, and spatterdock, with pondweed and other submergents in the openings. In the *regularly flooded salt marshes*, cord-grasses appear on the Atlantic and Gulf coasts, and alkali bulrush, glasswort, and arrow grass on the Pacific. *Irregularly flooded salt marshes* are dominated by needlerushes.

Bogs in northern regions support a spongy cover of mosses, and the typical plants are heath shrubs, sphagnum, and sedges. The southern boglike habitats, often called pocosins, bays, and savannahs, support plants such as cyrilla, persea, gordonia, sweet bay, pond pine, and pitcher plants.

In addition to the mangrove swamps, two other types of swamps may be distinguished: shrub and wooded. *Shrub swamp* vegetation includes willows, alders, buttonbush, dogwoods, and swamp privet. *Wooded swamps* in the North support larch, arborvitae, black spruce, balsam, red maple, and black ash. In the South, water oak, overcup oak, tupelo gum, and cypress are the typical trees. Western hemlock, red alder, and willow dominate in the Northwest.

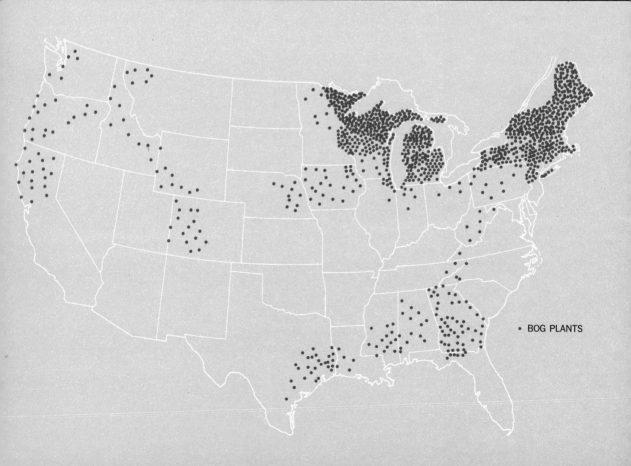

• BOG PLANTS

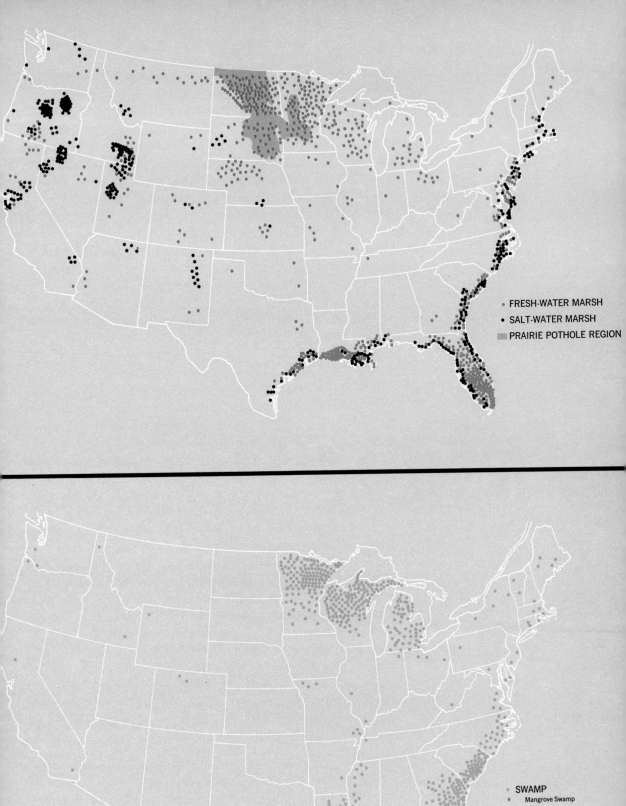

FRESH-WATER MARSH
SALT-WATER MARSH
PRAIRIE POTHOLE REGION

SWAMP
Mangrove Swamp

insects while trying to keep from falling up to your neck in treacherous mud.

More than seventy million acres in the United States, excluding Alaska and Hawaii, are covered with bogs, swamps, or marshes which are wet either permanently or during only part of the year. They support an almost unbelievable quantity and variety of wildlife: fourteen-foot alligators in Florida, the exceedingly rare whooping crane in Texas, countless thousands of muskrats in Iowa marshes, millions upon millions of ducks in the prairie potholes of Minnesota and the Dakotas, and so on, from one end of the country to the other. You could compile an ever-enlarging list of animals living in the constantly fluctuating environment of the wetlands.

Wetlands unlimited

Some splendid examples of wetlands are surrounded by the steel and concrete of great cities. One of the greatest is New York City's Jamaica Bay, a famous marsh containing three thousand acres of water and green flatlands, a haven for the ducks, geese, and shore birds migrating along the Atlantic flyway.

Other famous wetlands dot the vast expanses of rural America. There are sometimes hundreds of prairie potholes, actually miniature marshes, within a square mile; together they constitute one of the principal breeding places for North American ducks. You certainly have heard of Florida's subtropical swamplands: Corkscrew Swamp, with its thick green forest of cypress trees, and Everglades National Park, containing one of the world's largest mangrove swamps. There are even salt marshes far from the sea, in Utah and Nevada, where remnants of vast salty inland seas support plants typical of coastal areas.

Each kind of wetland—bog, fresh-water marsh, salt-water marsh, swamp—has its own particular kinds of plants and animals. Let us take a closer look at the wetlands and see how they are formed and how they are changing.

Federal and state waterfowl administrators distinguish four flyways for migratory birds in North America: Atlantic, Mississippi, central, and Pacific. The migrants do not follow these courses exactly—some even switch flyways during their journeys —but the flyway concept, based on the drainage pattern of North America, is useful to administrators.

Corkscrew Swamp Sanctuary, Florida, contains part of the largest surviving stand of mature bald cypress in the United States. Standing as high as 130 feet, these trees are among the oldest in eastern North America, some dating back more than two centuries before Columbus.

CRISIS IN THE EVERGLADES

The Everglades is a broad flooded plain curving a hundred miles southward from Lake Okeechobee to the Gulf of Mexico on Florida's west coast. The southernmost part, the "natural everglades," has long been one of the most interesting preserved wetlands in the United States, Everglades National Park. Here fields of saw grass stretch into the distance, dotted with hammocks of hardwood trees that rise like islands from a green sea. Now and then the pattern is broken with dense forests of cypress or mangrove swamps interlaced with watercourses along which wildlife in an incredible variety flourishes. Since 1948, however, programs of water diversion for agriculture and for flood prevention have disturbed the natural balance so severely that the Everglades is in danger of drying up forever. The one-foot-high tide of fresh water that used to well forth from Lake Okeechobee each summer and fall kept the delicate balance intact. That water is now regulated to such an extent that until recently less than the minimum amount necessary for the survival of the park was released into the Everglades each year. While the political controversy over water for the Everglades goes on, much of its wildlife is being lost. A large portion of the once lush wetland is now parched and dry, ravaged by fires that roar unchecked through the saw grass. Unless action is taken soon, the Everglades may very well disappear forever.

A female alligator stands watch near her nest in a fresh-water slough. This unique reptile heaps up a mound of decaying vegetable matter and mud in which she deposits her eggs. The rotting debris generates heat that helps to incubate the eggs.

Many of the birds that live in the Everglades
depend in one way or another on the
watercourses flowing through the region.
The diversion of the natural flow from
Lake Okeechobee has imposed severe
hardships on many of the birds, and, for
one at least, may mean extinction. The
anhinga, or snakebird (opposite) used to
delight visitors along the Anhinga Trail
in the national park as it dried its enormous
wings in the tropical air. Only a few of
these adept fishing birds now remain in
the park. Just thirty years ago wood ibises
nested in Corkscrew Swamp and the
Everglades in numbers estimated at 100,000
birds. At least a million pounds of fish
were needed to support the nesting birds.
Because of the reduced water flow in the
area, the fish population has decreased to
such an extent that these sole North
American representatives of the stork
family now nest only in small isolated
colonies (below). The everglade kite
(right) faces the most serious threat,
because of its specialized feeding habit.
It eats only one kind of snail. Since drought
is killing the vegetation on which the snail
feeds, fewer snails are able to survive. As a
result, the everglade kite is in great
danger of extinction.

Flushed from an Everglades
watercourse, a roseate spoonbill
takes water with it as it explodes
into the air. This spectacular bird
sweeps its distinctive bill back and
forth through the water to sift
out the tiny crustaceans on which
it feeds.

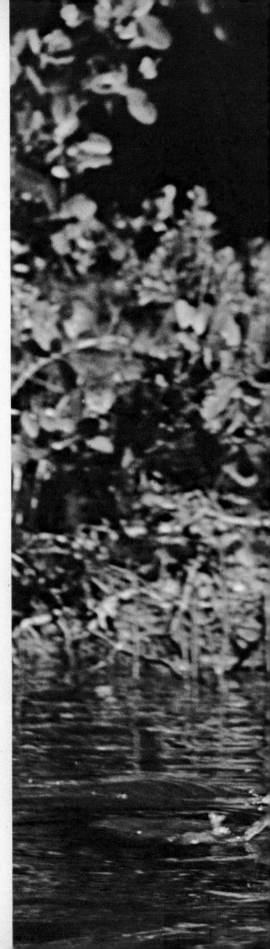

One of the Everglades' gaudiest
inhabitants is the purple gallinule.
This bird treads fearlessly out on
floating plants to feed on the flower
seeds and insects swarming on
the blossoms. Its widespread toes
help to keep it from slipping off
into the water.

The birth of a bog

Lakes that have relatively steep banks and poor drainage often become bogs. If you happen to live near a bog, you can see the dramatic changes that take place as the lake fills in.

Bogs usually form in regions that were covered by glaciers until about ten thousand years ago. Within this glaciated region many bogs have developed in what are called *kettle-hole lakes*, formed when huge ice blocks were buried in acres of sand, rock, gravel, and other debris ground up by the advancing ice sheet. Eventually the ice blocks melted, leaving deep, water-filled basins, usually with neither an inlet nor an outlet.

Far north in the tundra there are bogs filled with arctic plants. Farther south, in the coniferous forests of southern Canada, spruce bogs called *muskegs* cover large, poorly drained areas. Similar though much smaller bogs extend southward well into the northeastern, north-central, and western United States; in forests of broadleafed trees they stand out as dark green spots, especially in the fall, when the surrounding forests have turned red, brown, and gold. Near the southernmost border of the region once covered by glaciers, the spruce is supplanted by another conifer, Atlantic white cedar. This tree is quite different from the spruce; it has flattened scalelike needles instead of the long, narrow, pointed needles of the spruce, and it sometimes grows as tall as ninety feet. North of the glacial boundary the northern white cedar, or arborvitae, is sometimes found in bogs and may replace the spruce. But wherever bogs exist, there is always a unique roster of northern plant species.

You can recognize most bogs by their dense cover of knee-high evergreen shrubs growing in a moist light green moss known as *sphagnum*. Often the moss and shrubs, along with grasslike plants called *sedges*, form a floating mat if there is open water. You can also expect a deep accumulation of partly decomposed plants, or *peat*, under the living mat. Bogs often have dense growths of coniferous trees as well.

The part of North America that was covered by successive ice sheets is a region heavy with bogs formed in the depressions left by melting glacial ice. The shaded portion of the map shows, in chronological order, the four major glacial advances, named after the states in which their deposits are most in evidence: the Nebraskan, Kansan, Illinoian, and Wisconsin. The terminus of each succeeding drift stopped short of its predecessor's; consequently, the four sheets of glacial deposit overlap like the tiles of a roof.

WISCONSIN PERIOD   KANSAN PERIOD
ILLINOIAN PERIOD   NEBRASKAN PERIOD

Bogs usually develop in steep-sided, water-filled depressions of glacial origin. Characteristically, they have poor drainage and contain acid-rich peat deposits that accumulate in the lake bottom as dead sphagnum moss and other plants rain down from beneath the mat growing out over the lake.

20

Exploring a bog

With a little exploration you can find bogs in various stages of development, and by comparing these different stages you can reconstruct the steps by which a lake turns into a bog. In some bogs, for example, you will see a sizable lake surrounded by a dense mat of vegetation. In others there are only tiny areas of open water, called *eyes,* indicating that the lake has been almost completely transformed. In still other bogs, the plants have covered the open water, leaving no immediate evidence that a lake ever existed in the spot.

Let us explore a bog that still has a small lake surrounded by a mat of low evergreen shrubs. Wear old sneakers or, better still, boots, since a bog is soaking wet underfoot. Starting from the upland surrounding the bog, walk toward the center, approaching as close as possible to the edge of the open water. You will be rewarded by an experience you are not likely to forget.

Before you leave the dry land, notice the kinds of trees around you. South of the Canadian spruce–fir forest you will probably be surrounded by broadleaf trees—oaks, hickories, maples, beeches, birches. As you leave the dry upland, you may pass through a swampy belt of red maples. Now you have to jump from hummock to hummock to keep from

Trees and other plants growing in sphagnum bogs usually exhibit a conspicuous zonation, with bands of different kinds of plants growing at various distances from the center. Often there are no distinct breaks between the zones; each one merges gradually with the next. Even so, the overall pattern is obvious. Nearest the center is a floating mat of sphagnum and sedges. Where the mat is a little deeper, low evergreen shrubs such as leatherleaf establish a foothold. Still farther from the center are large coniferous trees, such as spruce and larch, often followed by a broadleaf swamp forest. Upland forest typical of the area grows on the dry land surrounding the bog. The bog shown here includes plants that might be found in northern New York or New England. Although many of the plants shown would be found in other areas, and the pattern of zonation — floating mat, shrubs, trees — would be consistent with that of most bogs, some plants might be different. For example, spruce trees might be replaced by cedars and a sphagnum mat almost entirely by sedges.

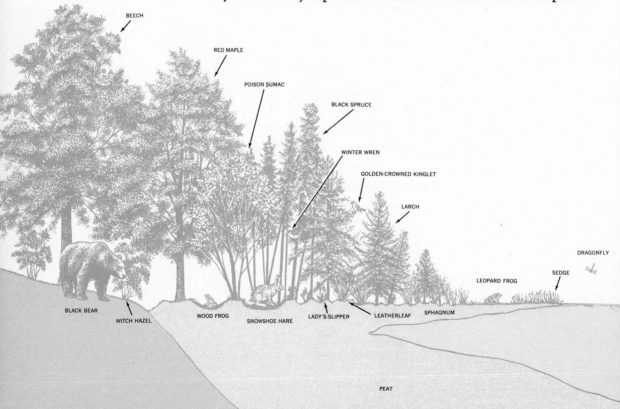

falling into deep peaty holes. Here is the place to watch out for poison sumac, a small tree with white fruits and leaves with several leaflets; like poison ivy, it can give you an uncomfortable rash if you come in contact with it.

Working your way deeper into the bog, you see your first black spruces, which may be scattered or else may form a belt surrounding the bog. As you pass through the spruces, bright green feathery-leaved larch trees, or tamaracks, catch your eye, and you begin to suspect that you are standing in the midst of a place different from any you have ever seen before.

A strange new land

Coming this far into the bog is like flying hundreds of miles farther north into a Canadian spruce muskeg within minutes. You half expect a moose to appear from behind a spruce tree at any moment! You are standing knee-deep in evergreen shrubs that spring from a carpet of spongy wet sphagnum moss. Here and there showy lady's-slippers send up their delicate flowers. You are in the heart of a northern bog.

Walk on among the scattered spruce and larch trees

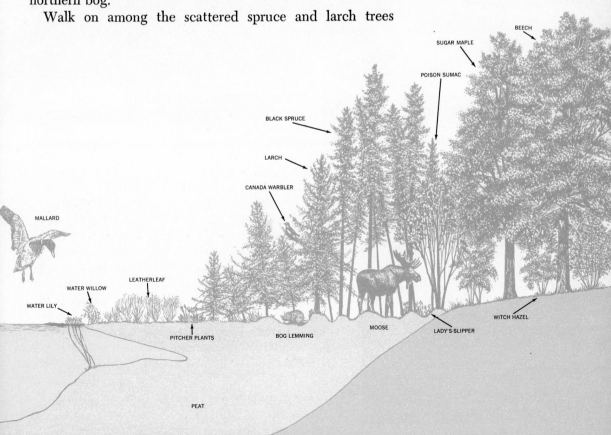

BEECH

SUGAR MAPLE

POISON SUMAC

BLACK SPRUCE

LARCH

CANADA WARBLER

MALLARD

WATER WILLOW

LEATHERLEAF

WATER LILY

WITCH HAZEL

MOOSE

LADY'S-SLIPPER

PITCHER PLANTS

BOG LEMMING

PEAT

Sphagnum, with its mass of minute leaves, is a typical bog moss. Alone or with other partly decayed plants it forms the thick bed of underlying peat characteristic of many bogs. Often sphagnum serves as a base on which other mosses such as haircap moss (*above*) can grow. The globular capsules near the center of the picture (*left*) contain dome-shaped clusters of spores which may, if conditions are favorable, eventually germinate, sending up new sphagnum plants. The leaves of sphagnum are a network of alternating large and small cells, which gives the plant a water-holding capacity of sixteen to eighteen times its weight, more than twice the capacity of cotton. In World War I sphagnum was often used in place of cotton as a surgical dressing, because it is naturally sterile.

toward the open water. If you stand too long in one spot, you may sink several inches or even a foot within minutes, because the sphagnum moss is so compressible. Approaching the open water, you notice that the mat you are walking on is no longer covered by shrubs. Now it consists of merely a thin layer of sphagnum or sedges, a floating mat that farther out is too thin to support your weight. Delicate water lilies may be seen out in the open water, and perhaps a few wild ducks paddling among the lily pads. The water lilies' stems are anchored in the peat beneath but send up their floating leaves on long stalks. Each year these leaves die and sink to the bottom, combining with the dead material falling from the underside of the mat to fill in the open water.

How is the floating mat growing out across the water? In some bogs a rather curious water plant known as water willow (not a true willow at all but a purple-flowering loosestrife) literally walks across the surface of the water. When the tips of its arching branches touch the water, they develop a spongy air-filled tissue and the stems float. From this spongy tissue many new arching branches form, creating an intertwining network of stems over the water surface, and within this network sphagnum moss grows and makes a floating carpet. Other plants find anchorage in this carpet, and so the bog grows. Don't step off the carpet, though, or you may go well over your head into the brown-tinted water beneath.

Loosestrife forms an arching network as it creeps over the watery eye of a bog. The buoyant air-filled tissue that develops on the submerged parts of the plant may swell the underwater stems to as much as four times the thickness of those abovewater.

A natural mummy

What would happen if someone did slip into a bog and disappeared forever? In Denmark, a man who had been purposely buried in the Tollund bog (presumably as a sacrifice to the gods) was found perfectly preserved after two thousand years. His facial features were sharp; even his whiskers were intact. Why such perfect preservation? In the bog lake, especially at the lower depths, there is very little life, since there is little, if any, oxygen. You will already have noticed that even the surface water has a rather brownish color, stained by the acids from accumulating peat. The presence of these acids, in conjunction with the low oxygen content, prevents many organisms from living in bog water. Bacteria and other microscopic forms of life, when they do exist in bog water, have difficulty breaking down plant

AN ANCIENT HANGING

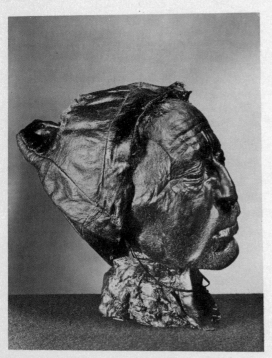

In May, 1950, a group of Danish peat cutters notified the police that they had discovered a body in a peat cut nearly seven feet below the surface of the Tollund bog in central Jutland. At first, the police believed that the corpse might be connected with a recent unsolved crime. Instead, scientists discovered that the body, whose skin had been stained a dark brown color like the peat that surrounded it, was the perfectly preserved corpse of a man who had been hanged and buried about two thousand years before. The man wore only the braided leather rope that had killed him, a peaked leather cap, and a belt. Probably he was a willing sacrifice to the goddess of fertility. About one hundred such preserved bodies have been found in bogs where the low oxygen content and the acids inhibit the microorganisms that cause decay. The head of Tollund man (*left*) is considered the finest preserved specimen, if only because it retains a remarkable vitality of facial expression.

and animal tissue without oxygen. The Tollund man was therefore almost perfectly preserved in the Danish bog; and it is because of similar conditions that our lake is gradually being filled in with peat.

LEATHERLEAF

Shrubs in the bog

On the return trip take a look at the shrubs about you. Near the water's edge they grow outward into the sphagnum or sedgy mat, helping to make it thicker—sturdy enough to support your weight. Sometimes shrubs or even sedges themselves form the advancing mat. Jump up and down on the mat and the trees wobble twenty-five feet or more away. This is a *quaking bog*. Even the pressure of your footsteps sets the whole bog shaking. The mat is floating on water that usually extends a considerable distance back under it.

How many different kinds of shrubby plants are there? Creeping over the sphagnum may be the tiny leaves and red fruit of the wild cranberry. (Sample the fruit, which is similar to the cranberries eaten at Thanksgiving and Christmas. In fact, domestic cranberries were developed from these wild ones.) A taller common shrub, leatherleaf, has smooth, leathery leaves, as its name implies. Labrador tea is also easy to identify by the thick tan fuzz covering the undersides of its leaves. The leaves of bog rosemary are silvery blue, and those of bog laurel are green above and white underneath.

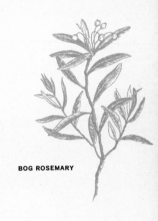

BOG ROSEMARY

The leaves of most of these shrubs are thick and leathery and have turned-under edges. This type of leaf is usually associated with plants of dry areas rather than bogs and swamps. In fact, sheep laurel, often found here, is also quite common in dry upland woods. Even with all this water, the bog is really a dry place for such plants, since they may have trouble getting water into their roots. How can this be true in a wetland? Perhaps the high acidity resulting from the presence of tons of peat or the accumulation of harmful chemicals in the stagnant water interferes in some way with the absorption of water by these plants. Whatever the reason, some bog plants have evolved the same general features as plants living in drier places. Above the water, the turned-under margins of the leaves shield the openings, or *stomates*, through which gases pass in and out

LABRADOR TEA

27

Not all the animal life in pitcher plants is dead. Among the creatures that regularly live in pitcher-plant water are the larvae and pupae of a small mosquito, Wyeomyia smithii.

PLANTS THAT FEED ON ANIMALS

Carnivorous plants are among the most bizarre of all living things. Yet, strange as it might seem, their adaptations for feeding on the bodies of small insects may have real value in the nutrient-poor soils in which these plants grow. All carnivorous plants—the pitcher plant, sundew, Venus's-flytrap, and butterwort—have poorly developed root systems. They characteristically grow in bogs or boggy areas. Although they are green plants and manufacture food by photosynthesis as other green plants do, it is believed that they obtain such important nutrients as nitrogen from the bodies of the unwary insects they lure, mire, or drown, and then devour.

Adrift in the watery reservoir of a pitcher plant, the corpse of a ladybird beetle (right) undergoes the slow process of digestion as strong enzymes break down all but its chitinous exoskeleton. The bright pitchers grow in clusters (left) or singly. The stiff downward-pointing hairs on the trap leaf direct hapless insects in the direction of a slick unhaired surface below, where they fall into the water.

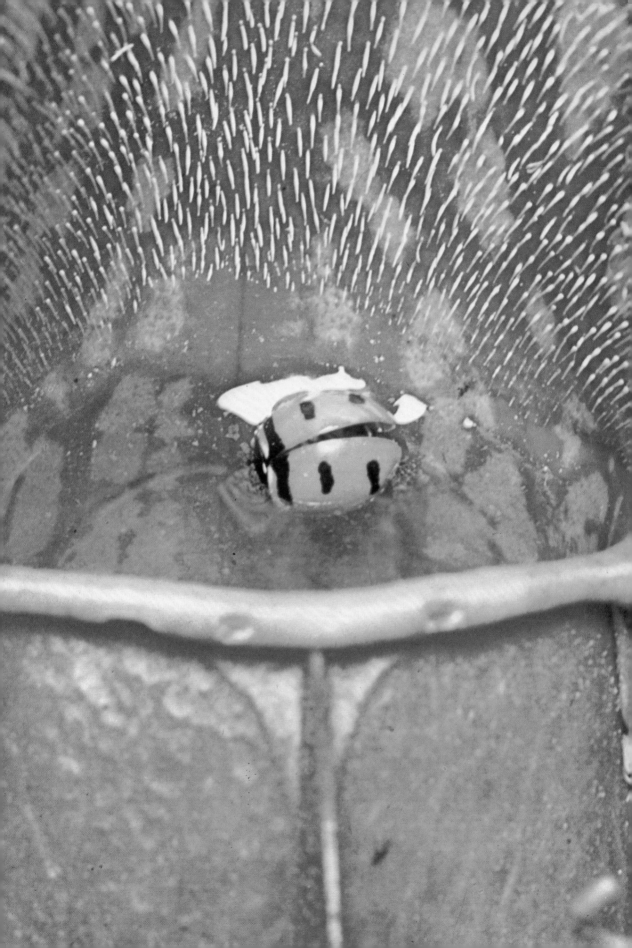

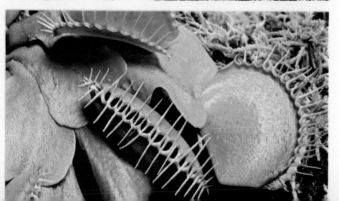

A sudden snap of the leaf imprisons the victim of Venus's-flytrap, one of the most dramatic of carnivorous plants, a native of a small area of the coastal plain of the Carolinas. At the left, a fly has landed, lured by both the color of the leaf's inner side and a substance secreted on its outer edge. The fly has touched two of the trigger hairs, or one of them twice, thus setting off the mechanism that closes the trap. If the insect is small enough to slip out between the teeth, it escapes. If not, it struggles helplessly until digestive juices engulf it. In about ten days the plant will have assimilated the nitrogenous salts of the body, after which the trap will open and the remains will weather away.

The butterwort poses no danger to a larger insect, and even ants and fruit flies can usually manage to free themselves from the sticky leaves that surround the base of the plant. The smallest of insects are its victims. One reason for the plant's ability to trap only small insects might be a protective mechanism: If a large dead insect is placed on a butterwort leaf, the leaf begins to decay. Charles Darwin studied the butterwort and found that the edges of the leaves curl inward after the insect is mired on the sticky surfaces. This movement has nothing to do with the capture of the insect but probably brings it in contact with more of the digestive glands.

Each yellow-green leaf in the two-inch rosette at the base of the butterwort plant is covered with two kinds of glands. Stalked glands produce the glue that pins down small insects; sessile glands not only produce digestive enzymes but also absorb the digested materials.

As a fly (right) struggles to escape from the glistening jewellike tentacles of a sundew plant, more and more of the sticky red hairs on the disk-shaped leaf will come in contact with the insect's body. Eventually the leaf completely enfolds the fly and the digestible substances are extracted from its body. Then the leaf opens and the remains fall to the ground. The action is slow compared with the sudden snap of Venus's-flytrap, but just as effective.

The threadleaf sundew is one of several species of this carnivorous plant, which grows in boggy areas throughout the world.

Ensnared on the long strands of the threadleaf sundew, the body of a dragonfly is digested in exactly the same way as prey caught on the bulging disklike tips of other species.

during photosynthesis. The curled margin helps prevent excessive loss of water from the plant by reducing the area exposed to drying air currents. The leathery coating serves much the same purpose as it does for desert plants by helping to prevent the loss of water.

The scene changes

As you move from the center of the bog toward its edge, you can expect to see more spruce and larch trees. Around the open water they often mingle, but sometimes they form belts—the larch first, then the spruce farther from the water. I can recall looking across Lake Itasca in Itasca State Park in Minnesota and seeing the light green larch extending over the sedge mat with the spruce beyond.

Why these distinctive bands of trees? Larch requires much more light than does spruce and thus grows best in open sunlight. Since spruce can flourish in areas that are more shaded, it can grow under larch trees and may some- day crowd out the larch. In some bogs in which the plants have completely filled in the lake, a rather dense spruce forest may develop with no trace of the larch that may once have stood there.

See if you can figure out the interaction of the trees in relation to the other plants in the bog you visit. Are the trees encroaching on the bog mat and shading out the sedges, leatherleaf, and other shrubs that grow best in sun- light, or is it a draw at the moment, with each holding its own? Just what is happening and how fast? When does it all stop? Will the spruce forest in turn be crowded out by the surrounding trees on the dry upland? This is not likely, since the peaty soil is probably not very suitable for the growth of upland trees. Some swamp trees, like those you

The filling of a lake in northern Michigan (*opposite*) has led to the development of a bog. As the bog progressed, distinct bands of vegetation were created. On the ground in another bog (*left*) the plant zones can be clearly seen. From the pitcher plant in the foreground the heath mat grades into a zone of bluish bog rosemary and then taller shrubs. But the trees in the background are deciduous ones, not the black spruces or other conifers you might expect. A zone has been skipped, and the pattern appears topsy-turvy to someone who thinks all bogs develop the same way.

Head down in a boggy pool, a bull moose, giant of the North American deer family, searches for food. The antlers of this bull span nearly six feet. He forages hour after hour in the ooze, submerging for as much as two minutes at a time . . .

saw when you started your hike into the bog, may grow there as the bog becomes drier. In some northern bogs *succession*, or vegetational change, may end with a spruce forest, because spruce-tree seedlings can grow up under the shade of their parents and replace any tree that is blown down or dies. Unless the environment changes radically, such a spruce forest may remain intact for a long time.

Animals of the bog

Many animals visit bogs, but few live there permanently. In Minnesota two ecologists, William H. Marshall and Murray F. Buell, found that different kinds of frogs lived on the sedge mat and in the bog spruce forest. Leopard frogs were abundant on the open sedgy mat, wood frogs in the spruce belts. The leopard frog can tolerate full sunlight and likes open water nearby to escape into. The wood frog prefers the cooler, more shaded woods where there are fallen logs or brush piles to hide under. You

36

might predict that the leopard frogs will decrease and the wood frogs will increase as the lake continues to fill in. This is one of the many kinds of changes occurring in animal populations as bog vegetation changes.

Keep an eye open for Muhlenberg's bog turtle, with its bright orange ear patch; man's destruction of bogs has made this species rather rare. In the southern white cedar bogs from the pine barrens of New Jersey southward, Anderson's tree frog, our most beautiful tree frog, can be recognized by its loud calls. Visit the Penn State Forest in New Jersey some June evening and you will hear its striking cry resound over the pine barrens.

Northern spruce bogs are fascinating places for bird watching. All around the rim of a bog, yellow-bellied fly-catchers, winter wrens, and Canada warblers dart among the branches of spruces and tamaracks. Overhead, in the spruces, tamaracks, and maples, red-breasted nuthatches, golden-crowned kinglets, and wood warblers in amazing variety—Blackburnian, Nashville, parula, magnolia—fill the air with song. Lincoln's sparrows search for seeds among the shrubby vegetation. Out on the thinnest part of the sphagnum mat the northern waterthrush, a wood warbler that seems to

... to nourish his fifteen hundred pounds on such seemingly dainty delicacies as pondweed and water lilies. When not eating, he might stand neck-deep in the water to escape the bites of the black fly.

BOGS — NATURE'S TIME CAPSULES

When the last of the glaciers retreated and peat slowly accumulated in the water-filled depressions left in the earth by huge blocks of melting ice, bogs began to record changes in the climate and landscape of the surrounding area. Pollen grains from trees growing around these bog lakes, carried by spring winds, sank to the bottom and mired there. As the conditions of the environment changed, so did the composition of the sediment on the lake bottom. Today, this thousands-of-years-old record can be studied by botanists to determine what kinds of forests existed at various stages in a bog's development and, consequently, what climate prevailed during each period. Botanists do this by sinking a hollow cylinder into the bog until a continuous sample from top to bottom is obtained. Then by studying the distinctively shaped pollen grains of various trees and noting their relative abundance at various depths in the sample, they can assign "layers" corresponding to different vegetation and climatic conditions. Firs and spruces, for example, grow in a cold, moist climate (layer 1); certain oaks and hickories, in a warm, dry one (layer 4); and so on. Sometimes a botanist will also find a black layer in the peat sample, the scorched record of a devastating fire. This bog represents a fairly typical history, though the record will vary from place to place. Samples from an Adirondack bog, for instance, will contain the pollen of trees somewhat different from those recorded in the peat of western bogs.

Peat samples from bogs not only reveal the history of the area in terms of its ancient forests and climates, but also can give quite accurate dates for those forests and climates. This information is obtained by a method called radiocarbon dating. After years of research this dating method became practicable in 1959 and is now widely used to date events occurring less than about 30,000 years ago. Essentially the process is this: Radioactive carbon, or carbon 14, is formed by cosmic rays as they bombard the earth's atmosphere. Once formed, the carbon 14 begins to emit beta rays and disintegrates at a fixed rate. In about 5570 years, half of its radioactivity is lost. Another half of the remaining radioactivity is lost in the next 5570 years, and so on. This loss is known as the half-life of carbon 14.

While a plant or animal is alive, its carbon 14 content is in equilibrium with the atmosphere and water, since carbon 14 moves from one medium to another in life processes. But when a plant or animal dies, its carbon 14 begins to disintegrate. Each 5570 years it loses half of its radioactivity. By a complicated process the radioactivity can be measured and the organic sample, in this case peat, can be dated. In this way pollen samples from the bottom of southern New England bogs, for example, have yielded dates for spruce forests some 11,000 to 13,000 years old.

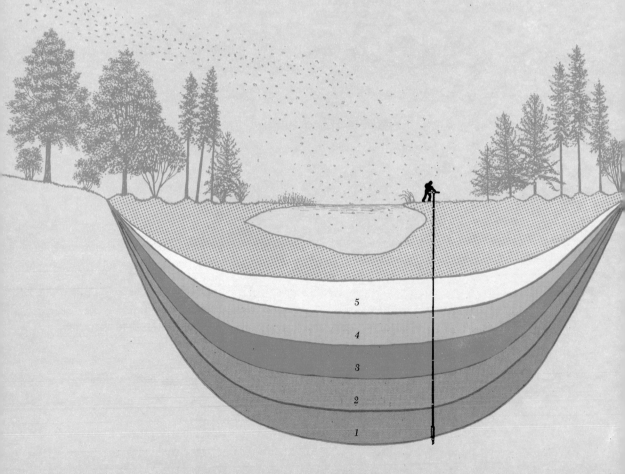

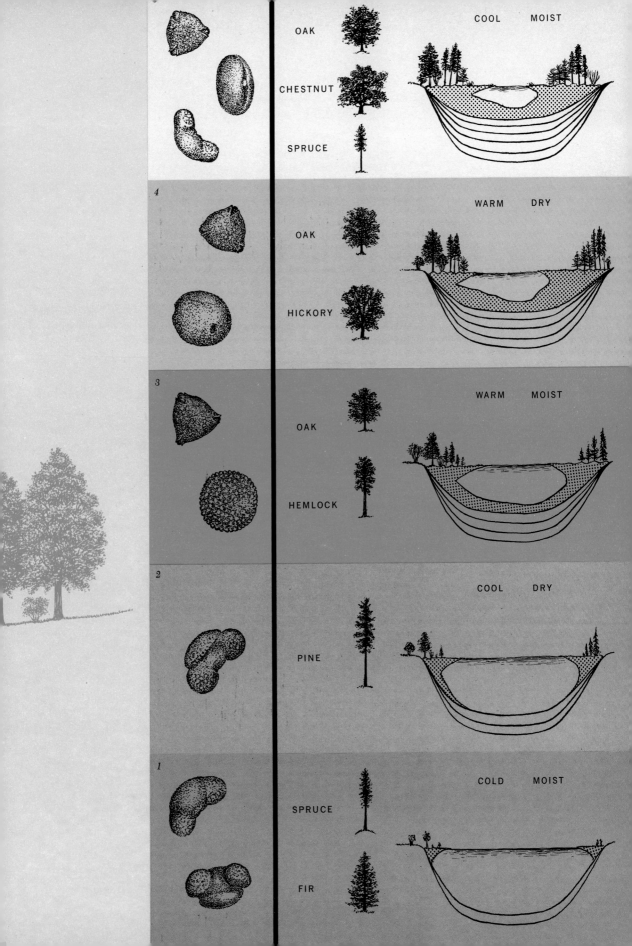

COOL   MOIST

OAK

CHESTNUT

SPRUCE

4

WARM   DRY

OAK

HICKORY

3

WARM   MOIST

OAK

HEMLOCK

2

COOL   DRY

PINE

1

COLD   MOIST

SPRUCE

FIR

be a cross between a sandpiper and a water ouzel in its habits, searches along the water's edge for mosquito larvae, caterpillars, and other insects.

Beavers, minks, and muskrats may also live here. In fact, the beavers may have built a dam across a bog outlet. Their presence can readily be detected by conically gnawed stumps and scattered felled trees not yet pulled into the water. Bears, too, make their rounds through the bog. Hunters occasionally chase one out during the late-fall hunting season.

The snowshoe hare is one of the most interesting animals you may see in a northern bog. Because the hare feeds largely on conifers, a bog is an ideal habitat for it. The fur of the snowshoe hare is brown above and white below in summer, all white in winter. An animal typical of the snowy north woods, the hare is protected from its many enemies by the excellent camouflage its white winter color provides. Large, long, furred, snowshoelike hind feet permit it to travel rapidly across deep snow.

Look for the signs of the small bog lemming—short-cut grass stems and little piles of green droppings in the lemming runways under the matted bog vegetation. Some bogs harbor isolated populations of these interesting animals, perhaps separated from the rest of their fellows as the glacial ice retreated northward ten thousand or more years ago.

Northern plants in the bog

Why are certain northern plants and animals found in bogs? One answer is that bogs are cool and moist much of the year. Spring comes late there because winter ice melts

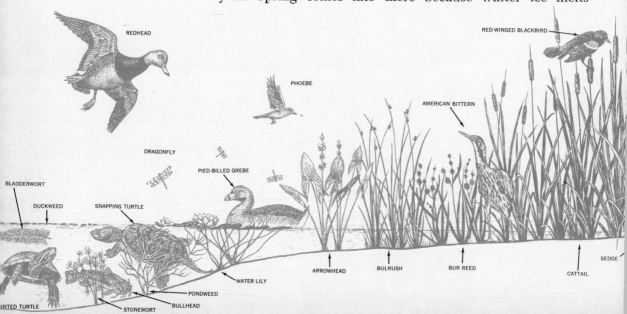

REDHEAD

RED-WINGED BLACKBIRD

PHOEBE

AMERICAN BITTERN

DRAGONFLY

BLADDERWORT

DUCKWEED

PIED-BILLED GREBE

SNAPPING TURTLE

ARROWHEAD   BULRUSH   BUR REED   SEDGE

WATER LILY

PONDWEED

INTED TURTLE   STONEWORT   BULLHEAD   CATTAIL

slowly under the insulated mat of vegetation. During the hot summer the wet sphagnum carpet acts like a huge sponge, evaporating great quantities of water and thereby cooling the bog considerably. In contrast to the surrounding upland it is a cool, moist habitat that tends to favor these northern plants and animals.

There are other factors in addition to climate. For example, do you remember seeing any really typical upland trees or shrubs in the bog? Since very few can tolerate the boggy conditions, even red maple, a typical swamp tree, often cannot thrive well in this stagnant, highly acidic environment.

All in all, the bog is a harsh place for ordinary terrestrial plants to live. You already know that some plants may have trouble utilizing bog water and that the minerals needed for growth are scarce, since bacterial activity is greatly reduced. Black spruce trees have measured only one inch in diameter at ground level, although they were more than thirty years old by ring count. Probably the surrounding plants that would normally penetrate a bog simply cannot stand these harsh conditions. For these reasons bogs are usually dominated by a distinctive and rather limited number of plants and animals.

The pond fills, and a marsh is born

Ponds, like lakes, are doomed to disappear. Just as a deep glacial lake may change into a bog, so shallow ponds and backwaters of river margins may be transformed into marshes. It is interesting to visit a shallow pond or river's edge where this change is taking place.

In the very shallow water along the edge of the pond,

Fresh-water wetlands may support several types of plant cover, the nature of which depends upon the depth of water, the climate, the type of soil, and other physical factors. Each kind of plant community has its own characteristic species of plants and animals. Plants that have their roots and lower stems in water grow in shallow water, and farther from shore floating and submerged plants can be found. Elsewhere there are sedge meadows that are periodically flooded, with small shrubs such as willows, as well as colorful wild flowers, growing on hummocks that rise like islands above the shallow water. In less flooded areas, thickets of pussy willow or alder many become established. The shrubs may grade into an occasionally flooded forest of water-tolerant trees, forming a wooded swamp.

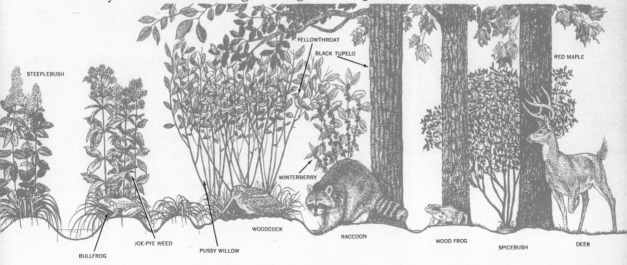

STEEPLEBUSH

YELLOWTHROAT

BLACK TUPELO

RED MAPLE

WINTERBERRY

WOODCOCK

RACCOON

WOOD FROG

SPICEBUSH

DEER

BULLFROG

JOE-PYE WEED

PUSSY WILLOW

cattails and bulrushes are able to get *rooted* in the mud, in
contrast to the *floating* mat so typical of the development
of a bog where the water is deeper. Cattails and bulrushes grow outward from the bank; but where the water is
deeper, water lilies take their place. If you have a canoe,
you can push out beyond the lily pads. The water is crowded
with plants, some rooted, others free-floating.

Scoop up one of the tiny bright yellow or purple flowers
poking above the water. It is a bladderwort, with narrow
leaves covered with little sacs. Some of the sacs appear clear,
but others are darker, as if filled with something. Under the
microscope they are seen to be traps that catch minute
aquatic animals, and sometimes you can actually see a tiny
water flea turning around inside.

The bladderwort is a kind of carnivorous plant, differing
from the others you have examined. By pressing gently on
one of the darker sacs while it is under the microscope you
can force out some of its prey, tiny animals such as water
fleas and mosquito larvae and tiny algal forms such as diatoms
and desmids. From the mouth of the bladder extend several
bristlelike hairs; these act as a trigger, setting off the deadly
trap when touched by a tiny passing organism.

How does the trap work? When the door is closed and
tightly sealed, tiny sets of four cells each, called *quadrifids*,
scattered over the inner wall of the sac, take up the water
in the bladder. This creates a partial vacuum that causes
the walls of the sac to become slightly indented—the same
thing that takes place when you suck in your cheeks.
Within half an hour after being placed in water, the traps
are set. Now picture a water flea swimming by and acci-
dentally touching the triggering hairs. The door of the trap
suddenly opens for a quarter of a second, and with a swish
the flea is sucked in as water refills the partly empty sac.
This is one of nature's most remarkable aquatic traps.

Submerged plants such as bladderwort thrive best in open
water where there is plenty of sunlight. As the lily pads

Low islands of vegetation dot an inland fresh-water
marsh in Montana. Fresh-water marshlands are treeless
expanses, often with dense growths of herbaceous plants
such as cattails, grasses, and sedges. In marsh pools,
where the water is deeper, water lilies, pondweeds, and
other plants become established.

As a pond fills in, the vegetation slowly converges on the remaining open water. Water lilies choke the water surface, and bulrushes, sedges, and other emergents reach steadily inward from the edge of the pond. As the vegetation changes and the pond becomes increasingly shallow, the animals of the habitat change too. Pond life disappears, and marsh life moves in.

continue to move outward, shading more and more of the water surface, most of the bladderworts, pondweeds, and other underwater plants will die; their leafy remains will sink to the bottom of the pond and contribute to the filling process.

So long as there is open water, the water lilies, with their elegant white or yellow blossoms, will persist. Turn over a lily pad. Its undersurface is a one-story apartment inhabited by aquatic life: tiny sponges, water mites, snail eggs, beetle eggs, damsel-fly eggs, and maybe a dozen more. The leaf-eating beetle does not even bother to go underwater to lay its eggs but merely cuts a hole in the leaf from the upper side and then lays them on the undersurface.

With further filling the water becomes more shallow and the parade continues: cattail, pickerelweed, bur reed, bulrush, buttonbush, water willow, and other plants continue to grow outward all around the pond. In some ponds fluctuating water levels or other factors may make it appear that the belts of vegetation are changing when they really are not. To be sure the pond is being filled in steadily, you would have to watch for many years.

You will remember water willow from the bog by its arching branches that literally walk across the open water. The buttonbush has a ball of flowers that become round clusters of fruit resembling buttons. Water lilies, pondweeds, and bladderwort now grow only in the remaining open deeper pools.

Red-winged blackbirds nest in the cattails. Dragonflies zoom all about. Ducks are dabbling in the shallow water or nesting on the banks, and herons are stalking their prey. Muskrats are moving in. This is a marsh, a treeless tract of water and aquatic plants, and it contains a tremendous abundance of life.

The death of a marsh

The marsh may live on, changing only very slowly if the water level remains high and there is little filling in by dead plants, but it may also die as a result of continued filling by silt and the remains of dead plants. As the water be-

Upended in a marshy pond, a female mallard searches for submerged aquatic plants such as pondweeds. The animal population of the marsh, as in any habitat, changes as the vegetation does. Until aquatic plants become established in the marsh, mallards cannot live there.

comes too shallow for floaters, bladderwort and water lilies, snapping turtles and bullheads may lose their place in the aquatic parade. The cattails, too, may pass away, and so may the long-billed marsh wrens and the red-winged blackbirds that nest among them. Some of the ducks will depart, and the muskrat population may decline. Many forms disappear because the changes in the habitat make it impossible for them to live there. But new ones will arrive to take their place.

Now that the declining marsh is less flooded, a beautiful sedgy meadow may develop. Clumps of grasslike sedge may rise as little islands from the muddy marsh like stepping stones; you can hop from one to another. With further filling an array of brightly colored plants may appear. You might see the lovely pinkish tints of steeplebush, joe-pye weed, or swamp milkweed; the rich purples of ironweed; or the bright gold of the wild lily and golden ragwort. It is summer now and, even though there are no pussies on the willows, you can still recognize this typical wetland shrub by its flattened buds covered with a single small caplike scale.

The bright red fruits of winterberry, a relative of the familiar Christmas holly, become conspicuous in the swamp as the weather turns colder and the green foliage disappears. This white-flowering shrub ranges up to fifteen feet in height.

The trees and shrubs move in

In the Northeast, trees and shrubs may already dot the marsh or form little thickets here and there. And now more of them begin to grow: maple, ash, elm, gum, blueberry, spicebush, and elderberry increase in the marsh. These and others may find a place in the filling of the marsh. In early spring you may see the soft yellow tinge of spicebush before the leaves appear. By midsummer the delicious highbush blueberries will be ripe. Later, elderberries can be gathered. The sweet smell that now penetrates the air may be sweet pepper bush, with its white candlelike blossoms, or swamp azalea, a relative of the showier ones planted around homes. The winterberry, as its name implies, shows off best near Christmastime. Its spectacular red fruits persist long after its leaves have fallen. Gallberry, or inkberry, a low evergreen holly with blackish fruits, also finds a spot, especially in swampy southern thickets.

Suddenly you hear a whistling sound and catch a glimpse of a stocky bird with a very long bill—the woodcock; he probably was resting, but in the evening he will be busy

46

probing for earthworms in the rich swampy soil. There is a flash of yellow, perhaps a goldfinch or a tiny yellowthroat. Catbirds, song sparrows, and wood thrushes are among the other songbirds that may find a nesting spot within the thicket's dense protective cover.

Now large trees such as maples and ash rise above the shrubs. But what has happened to our sedgy meadow? Too heavily shaded by the leafy branches of the shrubs and trees, the sedges, which flourish under the full sun, die off; the trees, growing still larger, form an upper story covering a deck of shrubs and small trees beneath.

The wooded swamp

On the ground the skunk cabbage may now spread its giant leaves, leaving little space for others. You may have missed its beautiful early spring flowers, which burst forth from the nearly frozen ground. They are able to bloom early because the heat generated by cellular respiration raises the temperature of the plant more than twenty degrees above that of the surrounding air. Nearby its relative jack-in-the-pulpit stands straight in its leafy covered pulpit,

Preacher Jack who stands in the jack-in-the-pulpit (left) is the club-shaped extension of the stalk, which bears the petalless flower of the plant. The hooded pulpit, called a spathe, is not a petal but the outside cover of the flower bud. In autumn, the spathe withers and discloses the gleaming red berries (above), its fruit.

but in the fall only its brilliant red fruits will mark where the plant once stood.

Under the leaves gray sow bugs that look like little pill-boxes scurry for cover. They break down the leaves and other litter and are therefore important soil makers. If you turn over a log or stone, you may expose the moist home of a red-backed or dusky salamander.

Some orange or yellowish jewellike flowers catch your eye. Hanging like little earrings, they produce a fruit that—if even touched when ripe—suddenly explodes, throwing its seeds far and wide. The plant is called jewelweed or touch-me-not. And there in an opening is the beautiful purple-fringed orchis, a flower to rival any from the florist.

A flash of red in the trees above—a scarlet tanager has now found a home. So have red-eyed vireos, chickadees, and nuthatches. Woodpeckers are sounding off as they drill the tree trunks for bark-inhabiting insects. Flycatchers patrol the air above the treetops. In the moss on the swamp-forest floor the hoofprints of deer indicate the presence of another visitor, and wood ducks are now nesting in a hollow tree nearby.

The marsh has become a swamp, a wetland covered with trees. Many forms of life have departed, but others have taken their place. From the shallow open pond you have seen a series of changes—one set of plants and animals replacing another—a kind of succession. Of course the change has taken place not overnight but after many years. You must be a keen observer and often study a wetland very carefully before you can discern just what is happening. Usually wet and occasionally flooded, this swamp forest will probably persist for a long time unless a colony of beavers moves in along the stream that flows through the swamp.

This forest was destroyed when beavers moved in and built a dam. The trees that were not cut for food or dam-building by the beavers have died, because they were unable to tolerate the high water level. A marsh is in the making.

The ocean water is rising

For thousands of years our eastern coastal marshes have been inundated at the rate of six to twelve inches each century—a five-foot rise in sea level every thousand years!

A thin sheet of water covers the soil in a Wisconsin swamp forest. Wooded swamps are covered with trees. Among the trees that have adapted to this environment are such species as silver maple, American elm, black willow, green ash, cottonwood, and river birch.

The spider lily grows in
fresh-water marshes at the
water's edge. Despite its
name it is not a lily but is
related to the amaryllis.

The pogonia, or beard-flower,
(above) is a member of the
orchid family and grows in
sphagnum bogs and along
wet, mossy shores in fresh-
water wetlands

The sea lavender is one of
the few wild flowers of salt
marshes. It often grows in
pannes, shallow depressions
that are highly saline.

WETLAND WILD FLOWERS

A variety of wild flowers grow in the wetlands,
but the special environmental conditions of bogs,
where soils may be alkaline or extremely acidic,
and of tidal marshes, where high salinity occurs,
prevent many flowering plants from growing.
Among the most interesting of the bog plants are
orchids. The small yellow lady's-slipper (*below*)
is one of them. The fringed gentian (*right*)
grows in wooded swamps and fresh-water marshy
areas. Its flowers open only in full sunlight.
On cloudy days and at night it closes into a
narrow bell shape.

How do we know this is happening, and what are its effects upon coastal marshes?

Huge southern white cedar logs buried ten feet deep were found by the New Jersey Highway Department when a highway across the Hackensack marshes was being constructed. These marshes now contain acres of *Phragmites*, the tall reed grass, but sometime in the past there must have been a cedar bog on this spot. Calvin J. Heusser, an ecologist who studied the area, found that the buried trees were as much as three feet in diameter and had lived to be more than three hundred years old when they were killed by the rising sea. As the dead trees fell, they were slowly covered by silt and by plants, such as *Phragmites*, that can grow in more salty water. As the sea rose, the cedar bog was buried by a developing coastal marsh.

The firmly fixed, interlocking roots of *Phragmites* crowd out other vegetation. Because these tall reeds are difficult to uproot, flood-control engineers can use them to help anchor the soil.

Coastal marshes in the making

At our new national seashore on Cape Cod, Massachusetts, you can actually see coastal marshes forming. In at least one of the inlets there was an open harbor used by sailing ships a few hundred years ago. Year after year silt has washed into the harbor, and large underwater growths of eelgrass also helped in the filling process. Today you can see dried masses of this same kind of grass along the beach, washed in by the tide. These inlets, or coastal bays, fill in just as lakes do, until small muddy islands appear at low tide and a tall-growing grass begins to take hold. Once started, salt-marsh cordgrass can spread rapidly by underground stems, forming scattered grassy islands. Eventually these islands may grow together, forming an extensive marsh covered by the tide much of the time.

Over the years silt collects among the grassy stems, and their dead remains keep building up the marsh just as they did in the bog. The marsh becomes flooded for shorter periods. The tall salt-marsh cordgrass dies out and is replaced by the smaller salt-meadow cordgrass, typical of

A watery meadow of cordgrasses dominates Barn Island marsh in Connecticut. Coastal marshes like this one are inundated regularly by nutrient-rich tidal waters that support countless forms of animal and plant life. In fact, they are among the most important of all natural habitats.

CHINCOTEAGUE NATIONAL WILDLIFE REFUGE

A vista of salt-water marsh grass stretches out west of Chincoteague Island, Virginia. The marshes are a section of one of the most extensive salt-water marsh areas in the continental United States. Teeming with wildlife, they extend from Maryland to the tip of Virginia's eastern shore. The Chincoteague Refuge includes the southern tip of Assateague Island, a thirty-three-mile barrier island and a national seashore that runs along the Maryland-Virginia coastline.

The American bittern points its bill and body skyward and blends into the background of salt-marsh reeds when an intruder approaches. It keeps its protectively striped side turned toward the source of danger, and if the wind sways the reeds, the bittern sways too.

Diamondback terrapins are among the few reptiles that live in salt-water marshes. They range from coastal waters well up into brackish estuaries but do not live in fresh-water areas.

The abundant wildlife owes much of its existence to the protection received from offshore barrier islands and to the twice-a-day influx of tidal waters. Atlantic brants, ducks, and clapper rails are among the thriving waterfowl and shore-bird population. Chesapeake Bay blue crabs and the famed Chincoteague oysters are also found here in great numbers.

The chunky seaside sparrow lives only in the salt marshes, where it wades along the exposed mudbanks to feed on small crabs and other animals. Its nest is found under shrubs, in patches of drift, and sometimes in thick grass stems, swaying safely above the tidal waters.

the higher marsh. There has been a succession—one kind of plant being replaced by another in the formation of the marsh. In this way our tidal marshes have been forming over the last several thousand years. Some have started in open shallow bays, others behind offshore bars.

Plant zones in the tidal marsh

If you look out over a tidal marsh, you can see different shades of green. The very dark green area is next to the upland, then there is a belt of yellow-green, and finally another shade along the water's edge—a kind of belting or zonation. Different kinds of plants respond to different salinities and degrees of flooding. Next to the water, as you would expect, is the taller salt-marsh cordgrass; and on the slightly higher marsh, the salt-meadow cordgrass. But the dark green highest up is different. It is *Juncus,* or black grass, not really a grass at all, but a rush related to the lilies. Its tiny flowers are like miniature lilies, though not so showy.

On the East Coast, these three belts often form the basic pattern from the water's edge to the upland. Near San Diego, California, on the West Coast, many of the same plants that grow in the East can be seen. However, the salt-marsh cordgrass, though closely related, is a different species. Unlike the pattern on the East Coast, there are often only two belts in western salt marshes—cordgrass near the water's edge and dense growths of glasswort in the higher marsh.

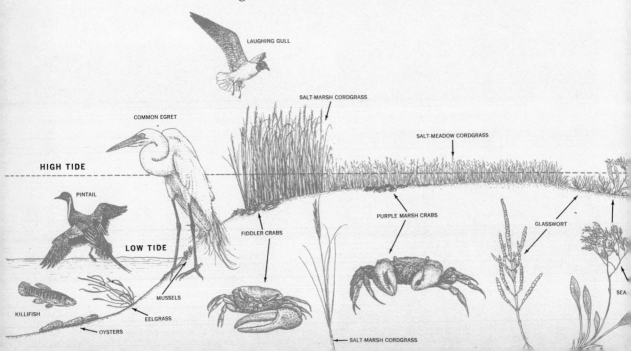

LAUGHING GULL

SALT-MARSH CORDGRASS

COMMON EGRET

SALT-MEADOW CORDGRASS

HIGH TIDE

PINTAIL

PURPLE MARSH CRABS

GLASSWORT

FIDDLER CRABS

LOW TIDE

MUSSELS

KILLIFISH

EELGRASS

SEA

OYSTERS

← SALT-MARSH CORDGRASS

Life in the ditches and salt pools

Out in the salt marsh, ducks are feeding in a pool a foot or two deep in the marsh. It is filled with ditch grass, a submerged aquatic plant similar to pondweed and an excellent duck food. To get to the pool you have to jump long mosquito ditches that cross the marsh. Fiddler crabs scamper into their burrows. Mussels are embedded in the exposed muddy banks. As you might expect, the grass along the ditches is different from that in drier areas: salt-marsh cordgrass now lines the ditch as far as you can see. Look at its flowers. Have you ever examined grass flowers closely? They are really very beautiful. Their tiny pale creamy blooms are clustered along one side of the flowering stalk. Those of salt-meadow cordgrass often have a purple tinge. Another variety you might see almost anywhere in the tidal marsh is spike grass, with its typical spire of flowers; you can also find it along the West Coast and in our inland salt marshes.

Jump another ditch. Try not to miss the solid ground on the other side, or you may lose your boot in the soft ooze. Here the pattern may change. It is wetter underfoot; salt-meadow cordgrass is doing poorly, and there is a shallow pool ahead in a low spot with no cordgrass at all. This basin, called a *panne*, is a shallow depression where high tide left a reserve of water to evaporate slowly. As the seawater evaporates, the panne may become so salty that the marsh grasses are unable to tolerate it.

But something is growing under the water. This is an algal mat, composed of blue-green algae that are distantly related to the green algae of the fresh-water marsh.

Salt marshes are covered by an array of plants that are alternately flooded by seawater and exposed to air by the rise and fall of the tide. As a result, marsh plants vary in their ability to withstand submergence. Each of the characteristic plant types is more or less restricted to areas where its living requirements are best fulfilled. Eelgrass grows only where it is covered by water at all times. At the edge of the bay and along channels where flooding is greatest, tall salt-marsh cordgrass forms solid stands. On slightly higher ground the shorter salt-meadow cordgrass covers vast areas. Black grass in turn grows best along the limits of the highest monthly tides, where flooding is brief and infrequent. Scattered throughout the marsh are shallow depressions, called pannes, that are flooded from time to time. The gradual evaporation of water makes the soil too salty for many plants to survive. Even so, pannes usually support sparse growths of stunted salt-marsh cordgrass, glasswort, sea lavender, gerardia, and a few other plants, including algae. Beyond the marsh, shrubs such as bayberry, holly, and others, depending on the area, grade into upland forests.

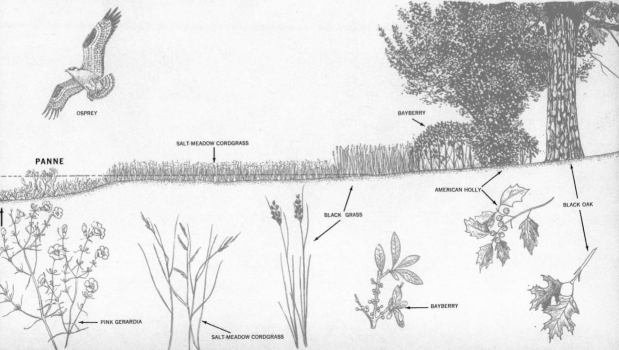

OSPREY

BAYBERRY

SALT-MEADOW CORDGRASS

PANNE

AMERICAN HOLLY

BLACK OAK

BLACK GRASS

BAYBERRY

PINK GERARDIA

SALT-MEADOW CORDGRASS

Farther back in the panne the ground is less wet. In the summer, flowers bloom here. Notice the sea lavender; even when its flowers fade, its dried flower stalks have a beauty of their own. Tiny white asters and also pink gerardias (relatives of snapdragons) add their color. The thick stems of glasswort form small dense forests in the bare spots. Green now, with tiny scalelike leaves, they will turn a brilliant red later in the season. The marsh is an array of color, quite a contrast to our first impression.

A slight change in elevation, an inch or two, or even less, and new plants appear. Ever-changing, this is the pattern of the marshes. Elsewhere along the Atlantic Coast the pattern will differ, but you can expect to find many of the same kinds of plants you have seen here in southern New England.

All kinds of wetlands

Each kind of wetland—whether it is a marsh, swamp, or bog, and whether its water is fresh, salty, or brackish—is a distinct *ecosystem*, a unique roster of plant and animal life contributing to the community as a whole. You have now briefly toured each kind of wetland and have seen some of their typical plants and animals. You have also observed how wetlands change, how lakes that are steep-sided can become bogs and how marshes develop from lakes that have gentle underwater slopes where floaters, submergents, and emergents can get a foothold.

But no plant or animal exists in a world of its own. Each form of life is inextricably tied to every other form of life. The survival of one living thing depends, directly or indirectly, on the survival of every other living organism in the community. Ultimately all life depends on the energy of the sun. The story of its integral role in the living community and of its incorporation into life-sustaining food is a fascinating one.

Black spruces, unable to become firmly anchored in the soft peat of a bog, lean at precarious angles amidst a carpet of ferns. The outer edge of a bog is a zone of light and shadow in the ever-changing world of the wetlands.

59

The Flow
of Energy

There is a pattern in nature, and we are very much part of it. Soil, water, plants, animals—including man—exist together in a system of interdependencies called a *food web,* which is a series of interacting *food chains.* For example, a simple food chain in Florida's Everglades consists of a plant-eating green snail and a rare bird, the everglade kite. The kite feeds only on this particular snail, nothing else. You could call these interdependencies a *plant–snail–everglade kite* food chain. However, the green snail feeds on several kinds of plants and is in turn food for other swamp birds, especially the limpkin. So it is a part of other food chains as well.

On a large chart of all the living things in a forest, desert, fresh-water marsh, or some other natural habitat, you could map out a food web. You would have to include a whole range of animals, including insects, birds, reptiles, amphibians, fishes—all are involved. By drawing lines between those living things that eat (or are eaten by) each other, you would soon have a maze of interconnecting lines running from every plant—the smallest bacterium and the largest tree—and every animal—the one-celled animals and the largest meat eaters. Some, such as the everglade kite, would

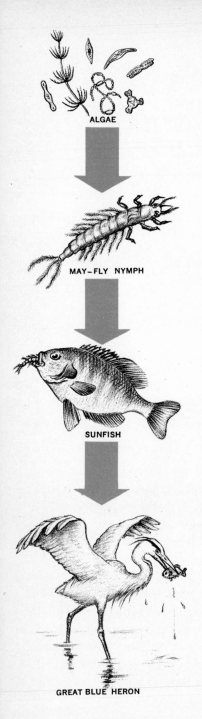

ALGAE

MAY–FLY NYMPH

SUNFISH

GREAT BLUE HERON

One of the most vital interrelationships of life is the food chain, a single pathway of energy from the sun to plants, here algae, and then to first-, second-, and third-order consumers: a May-fly nymph, a sunfish, and a great blue heron, respectively.

The bite of the inch-and-a-half-long fish spider sends a deadly venom running through the blood stream of a fish. To find its prey — fishes or insects — this spider walks on the water as gracefully as a water strider or submerges for as long as forty-five minutes.

have only a few lines, but most others would have many lines connecting them to hundreds of other plants and animals in the food web.

You can investigate a food web in the wetlands. For our purposes it is easiest to confine the study to two kinds of wetlands, the fresh-water marsh and the salt-water marsh. However, the delicate and fascinating food web you will discover in these two places is part of the same master pattern that exists everywhere in the living world.

Life and death among the cattails

You begin your study of a food web in a fresh-water marsh. Before you stretches a large wet tract of treeless land. As far as you can see, dense growths of cattails, bulrushes, and other plants create an enormous green carpet interrupted here and there by patches of open water where ducks are paddling about in search of pondweeds.

In a small pool a female mallard duck is followed by four ducklings trying desperately to keep pace with their mother as they take their first swim. Suddenly the last duckling in line disappears. A snapping turtle has snatched it from below.

In a moment all is quiet again. The sun shines on the still water, and the cattails sway in a light breeze. All activities go on as usual. A duckling has lost its life, and a turtle has gained the supply of energy it needs to stay alive. The commonplace death of a duckling has helped preserve the continuity of life in the marsh: the energy stored in the duckling's body has been transferred to another animal.

You will witness hundreds of such events while studying the marsh. Here, as everywhere else on earth, animals must eat to live. Life is activity, and activity requires energy at every moment. Swimming, flying, crawling, eating, breathing, and growing use up energy, and all animals must replace the energy they use by feeding on plants or other animals.

Sunlight, the ultimate source of all energy, falls on the spearlike green leaves of cattails. By the process called photosynthesis, tiny factories, or chloroplasts, in plant cells convert solar energy to chemical energy and store it in the form of food, which sustains all other forms of life. This conversion can occur only in those cells that contain chlorophyll, the green coloring matter in plants.

A link with the sun

The initial source of all energy is sunlight. On a clear day it pours down upon the earth and sea in a steady stream. The green plants that live on land and in the water contain substances known as *chlorophylls*, which in the presence of sunlight are able to use carbon dioxide from the air and hydrogen from water and, in a split second, turn them into a simple sugar called *glucose*, the basic food of life. Only green plants are able to store and utilize the sun's energy in this way. The process is called *photosynthesis*, which means "putting together with light."

But, since animals cannot capture and store solar energy in this way, they must get their energy by eating plants (as when ducklings eat arrowhead tubers) or by eating other animals (as when snapping turtles eat ducklings). Thus all the animals that inhabit any given place are bound together by their dependence on its plants. The plant eaters in such a place, or *habitat*, obtain their energy directly from the plants; the meat eaters obtain theirs by eating the plant eaters. Larger meat eaters in turn get their energy by eating

Every living organism uses the energy that comes from the sun and passes it on in a cycle. The energy that green plants, or producers, capture from sunlight is passed along to plant eater and then to meat eater. As energy is transmitted from one consumer to the next, some of it is lost in the form of heat, a by-product of metabolism. The remaining energy continues in the cycle from consumer to consumer, even supplying some energy to decomposers. Then matter, in the form of chemical compounds, is returned to nourish plants, and new energy captured from the sun continues the cycle.

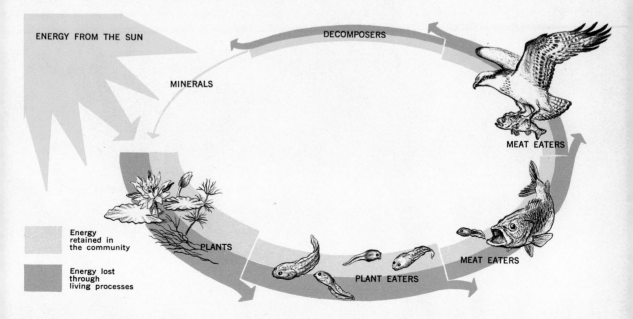

ENERGY FROM THE SUN

DECOMPOSERS

MINERALS

MEAT EATERS

Energy retained in the community

Energy lost through living processes

PLANTS

MEAT EATERS

PLANT EATERS

smaller plant eaters or meat eaters (as when foxes eat rabbits or hawks eat water snakes). Of course, some animals, like man, eat both plants and animals.

Because of their unique ability to manufacture energy-laden substances, green plants are called *producers*. All other living things, including certain nongreen plants such as the fungi, are *consumers*; they procure energy from plants or other animals. In forests and deserts, on mountain slopes, and under the sea, the story is the same. Vegetation may vary and animals may be of widely different species, but only the green vegetation makes food—with the help of solar energy.

Microscopic producers

By far the largest number of algae are microscopic plants. But despite their size, these plants are at the base of all food chains in the wetlands. Without these primary producers and the energy they capture from sunlight, many of the wetland animals could not survive.

The green filaments you see floating in the marsh water are green algae, very primitive plants. E. S. Barghoorn, a botanist at Harvard, has recently presented evidence, including photographs of structurally preserved fossil plants, showing that some kinds of algae had appeared on earth in the pre-Cambrian period, two billion years or more ago. They are a basic food source in all wetlands. Many insects, such as the water boatman, live almost exclusively on the food that green algae manufacture. Tadpoles, crustaceans,

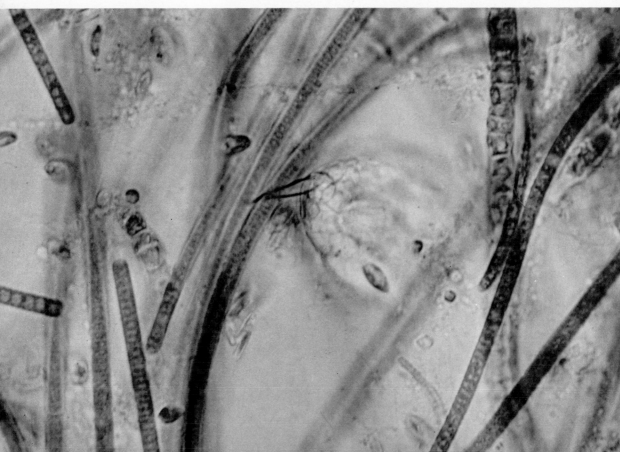

and fishes also feed on algae. If the algae were removed from this marsh, billions of microscopic animals as well as insects, tadpoles, and small fishes would starve, and so would the larger creatures that feed on these insects, tadpoles, and small fishes.

Algae have a simple structure: they lack the leaves, stems, and roots of many plants you know. Under a microscope they are usually seen as chains or clusters of nearly identical cells. Some algae, the giant seaweeds, grow to lengths as great as two hundred feet, but nearly all species are smaller; indeed, most are microscopic. You can see some of the smaller varieties floating in ponds and marshes, sometimes as a green scum on the water surface. But whatever their size or appearance, all algae are true plants, true producers of the energy-laden substances that support animal life.

Some other marsh plants

Cattails, which have clublike flowering stems, grow densely where the water is not too deep. Interspersed with the cattails are profuse growths of bulrushes. Bulrushes belong to a group of grasslike plants called *sedges*, many of which have triangular stems, as you will realize when you twirl one between your thumb and index finger. Sedges grow in all wetlands, whether marsh, swamp, or bog.

The long, slim plant with the blue flower is pickerelweed, another marsh dweller. Next to it you will find some clumps of arrowhead, which is named for its leaves shaped like spear points. If you pull an arrowhead out of the water, you will find several bulbs dangling on the stems. Since ducks regularly feed on these succulent bulbs, arrowhead is also called duck potato.

Floating on the water are the flat, broad leaves of the water lily. These are connected by long, slender leafstalks to tough stems, sometimes as thick as a man's wrist, which are buried in the muddy bottom.

Between the lily pads you can find thousands of perfectly formed minute plants that are not rooted to the bottom —they just float. These are duckweeds, one variety of which is the tiniest of all flowering plants.

Below the surface of some pools lies a whole world of submerged plants adapted to a completely underwater existence. One of the most common is the sago pondweed, an

BULRUSH (SEDGE)

WILD RICE (GRASS)

Both grasses and sedges are abundant in marshes. Most grasses, such as wild rice (*left*), have round jointed stems that are hollow between each joint. Most sedge stems are triangular, as you can discover for yourself by twirling one between your fingers.

The duckweeds are suspended floating plants that come in contact with soil only if they are blown ashore. Riding currents, they can cover great distances. Because they reproduce rapidly, in a few weeks they can cover the surface of a pond (*far right*). At the top of the close-up are the plants of lesser duckweed; at the bottom is water flaxseed. The tiny plants in the center, called watermeal, are the simplest and smallest of all known flowering plants in the world.

important food for ducks and other waterfowl. In spite of an unpromising appearance—stringy stem and long, narrow leaves—this sago pondweed and its many relatives contain millions of soft, nourishing cells. Their small fruits are especially liked by marsh birds.

Consumers large and small

At the edge of a patch of open water a clump of bulrushes stirs. Through your binoculars you can see the cause of this activity. A great blue heron is stalking prey; as you approach, it takes flight on large wings. You should have no trouble finding the big frog that was the intended victim, for it springs suddenly into the water with a startling splash.

You begin to realize that the whole marsh is stirring with life. Half-grown tadpoles dart to conceal themselves in the surface mud. A school of small fish flashes out of sight. In the water tiny aquatic insects and crawlers scurry across the bottom or hide in the greenish threadlike strands of algae.

The marsh water is in fact teeming with tiny organisms that feed on algae. Water fleas, or daphnia, are particularly abundant. These barely visible animals are related to crabs. Under their transparent shells you can see tiny digestive tracts, stained green by the algae they have eaten. Using a magnifying glass, you can observe the ten or twenty embryos whirling around in the brood chamber of each female. Once this brood has been born, it could, within a month, produce six thousand additional water fleas, but few spend much time at liberty before they are gobbled up by minnows.

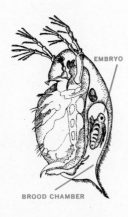

EMBRYO

BROOD CHAMBER

A tiny daphnia, or water flea—one-sixteenth of an inch long—carries its unborn young in a transparent pouch on its back. This prolific crustacean is a diet staple for fish and other marsh life. One kind of daphnia produces a brood of eggs every two days. In sixty days its descendants could number thirteen billion.

68

Fish—and the fisherman—are consumers too

Where the water is deeper, you can fish and gain for your-self some of the energy passing along the food chain. A flexible stick, a bit of string, a fishhook, and some earth-worms for bait are all you need. The line goes taut when something underwater seizes the bait and tries to swim away with it. The tugging at the line continues until you jerk it out of the water. Dangling at the end, still fighting, is an attractive eight-inch fish with a round body and a bright red spot on its gill cover. This is a pumpkinseed sunfish, one of the typical species in American ponds and marshes; two or three make a tasty lunch. If you prefer not to be the consumer, you can set the sunfish free to become a meal later on for a raccoon or a mink. Another fish you may catch is the little yellow perch, which travels in schools, feeding on plankton, insects, and other fish.

Other fish good for eating are found in the marsh. One of the most delicious is the black bullhead, a member of the catfish family. Most catfish have scaleless skin, and the eight projections, or *barbels*, around their mouths resemble

The pumpkinseed sunfish (*above*) abounds in the weedy waters of ponds and marshes and feeds on insects and small crustaceans. The carp (*below*) was introduced from Europe about 1800, and even though it sometimes muddies the water so much that plants cannot grow, it has proved commercially valuable.

a cat's whiskers. You are most likely to catch this species at dusk, when it becomes especially active. If you get a bullhead on your line, handle him carefully. Those little spines on the fin tips near the head are strong and sharp and can inflict nasty puncture wounds.

The bullheads in this marsh rarely exceed twelve inches in length, but they are close relatives of the huge channel catfish that live in the Mississippi River. Some there are more than five feet long and weigh a hundred pounds.

If there is a stretch of open water in the marsh, you may catch a bass, which likes to lurk around the lily pads, searching for aquatic animal life.

The variety of fishes in marsh water is limited. Only a few species can live there successfully because they must be able to tolerate both muddiness and warm temperature. Since such water holds relatively little oxygen, catfish and other marsh fishes must be able to get along at times with a very restricted oxygen supply. The large active fishes of deep water and open ocean would die of asphyxiation in the water of a marsh. Furthermore, during extreme drought some of the pools in a marsh may dry up, creating another hazard for the fishes.

The striped, or black, mullet lives in coastal waters and brackish estuaries along the Atlantic Coast as far north as Maine. As the fish grows, its intestine becomes longer and more convoluted, probably the result of a change in diet: the fry feeds on plankton, the adult on larger plants.

71

Animals in the marsh

Animals here, as everywhere, have two pressing concerns: to eat and to keep from being eaten. There among the duckweed is a green frog, its body hidden beneath the water; only the bulbous eyes are visible. The frog's tongue lashes out to seize a luckless insect. It gobbles up the prey so quickly that you do not have time to identify it. It may have been a damsel fly, or even a May fly that has just emerged from a nearby stream. Frogs will eat anything that hops, crawls, flies, or runs, provided the victim is not too large or too powerful. Frogs' tongues are attached to the front edge of the mouth so that they can unfold and be thrust out rapidly to full length. Small frogs eat insects most of the time; larger bullfrogs capture small fishes, snakes, mice, and ducklings and other young birds.

If you look sharply, you can see a water snake swimming smoothly through the shallows, clutching in its mouth a fish that looks like a minnow. Some people are appalled by snakes and always try to kill them. This is short-sighted. Snakes are beneficial. By eating numbers of small fishes such as perch and sunfish, they keep these populations from growing too large and crowding out themselves and other species, such as bass and pickerel, that are more attractive to fishermen. Snakes in woodlands and grassy country destroy millions of rats and mice every year. Even the poisonous snakes, such as water moccasins and massasaugas, that live in some American wetlands help keep the wildlife population in a proper balance and perform a useful role in the marsh economy. These snakes are shy creatures that do not attack unless molested.

Water snakes find marshes rich hunting grounds, but they are themselves always in danger from hawks and other predators. The snake you just saw has also been noticed by a marsh hawk. As the bird circles overhead, use your binoculars to study the long, slim wings and tail and the white rump. The marsh hawk nests on the ground in the reeds at

The water moccasin, or cottonmouth, is a dangerous poisonous snake, but it plays a valuable role by keeping in check many of the prolific species in marshes and swamps. Cottonmouths prey on fishes, frogs, salamanders, and other small forms of animal life.

The red-winged blackbird, one of the most common birds south of Canada, is a marsh dweller. At dawn it leaves its roosting place to feed, sometimes traveling as much as thirty-five to fifty miles a day. The male, with his brilliant epaulets, is the bright-plumaged sex....

the edge of the marsh, but his useful services are not confined to this habitat. He is also an inveterate hunter of meadow mice; one pair of marsh hawks will eat a thousand destructive mice in a single nesting season. Mice nibble on the bark of fruit trees and, by girdling their trunks, kill the trees. A concentration of fifty mice an acre can devastate a whole orchard. Marsh hawks and other predatory birds help keep down the mouse population and thus are an invaluable form of rodent control. For centuries men have shot hawks without thinking about the consequences, although we are now beginning to recognize the foolishness of such action. These predators help prevent population surpluses and in the long run protect our interests.

As you circle the marsh, you see that something has been feeding heavily on the plants that emerge from the water. Cattails, bulrushes, and other growing things have been nipped off. Here and there water lilies are floating free, deprived of their thick, nutrition-transporting stems. This is the work of muskrats. They are primarily plant eaters; emergent plants, supplemented by an occasional meal of mussels, make up their diet.

Here and there you find tracks, small catlike prints left by a mink. Unlike the muskrat, the mink is carnivorous, feeding on fishes, frogs, and occasional birds and, when other food is scarce, sometimes preying on muskrats.

Birds you can't miss

A shrill *konk-ka-reee* rises from the cattails, and a sleek red-winged blackbird soars out. You admire his shining jacket and the bright red patches tipped with yellow on his wings. By approaching slowly, you can get within sighting range of

The soft earth of the marsh records the movement of its animal inhabitants. A passing muskrat has left its tracks astraddle the trail of its tail. Its feet are five-toed, but the tiny inner toe of the smaller front foot, just visible here, seldom leaves its mark.

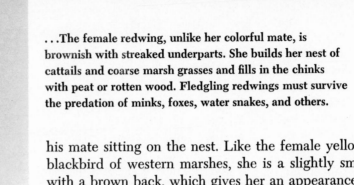

...The female redwing, unlike her colorful mate, is brownish with streaked underparts. She builds her nest of cattails and coarse marsh grasses and fills in the chinks with peat or rotten wood. Fledgling redwings must survive the predation of minks, foxes, water snakes, and others.

his mate sitting on the nest. Like the female yellow-headed blackbird of western marshes, she is a slightly smaller bird with a brown back, which gives her an appearance very different from the male's. It is hard to believe that the male and female belong to the same species.

Redwings are especially numerous in the prairie pothole marshes of the Midwest and, in winter, in the rice-producing wetlands of Louisiana and Arkansas. Some farmers consider them highly destructive to grain crops and are allowed to shoot them on sight when they do damage. Yet these sleek, vividly marked blackbirds probably do us much more good than harm. By analyzing the stomach contents of a thousand redwings, biologists found their eating habits quite different from what had been supposed. Insects, including such plant pests as weevils, constituted 30 percent of their meals. Weed seeds made up the bulk of their diet. They consume quantities of ragweed (its pollen causes a type of hay fever) and eat prodigious quantities of other weed seeds, thinning out troublesome plant species. Even in late summer, when corn, wheat, and oat crops are readily available, they concentrate mostly on seeds worthless to man. Only 13 percent of their diet consists of grain—a small price to pay for the work they do in weed and pest control.

A shrill little warble that sounds like *witchity-witchity* announces the presence nearby of a yellowthroat. This small bird, which lives on insects, nests on the ground in the damp shrubby thickets on the edge of the marsh. Even without seeing the color of the male's throat, you can identify him by the black mask across his eyes. If you hear his warble but cannot find him, you may coax him out of hiding by expelling your breath sharply through your mouth. The resulting *pssh-pssh* sound usually does the trick. Kissing your hand loudly has the same effect. Attracted by these sounds, the bird will fly close and study you while you study him. Mechanical squeakers for attracting birds are available also, but they should be used with caution. Someone who once used one in a Florida marsh found himself face to face with a hungry bobcat that came slinking out of a thicket!

Too many raccoons

Because of the water supply and the abundance of plants and animals to eat, the marsh is often visited by hungry creatures from the uplands. The raccoon frequently hunts in the shallows for fish and crayfish, flicking them into his mouth with one adept paw. Raccoons dig up whole batches of turtle eggs buried in the sandy banks, and they raid duck nests to suck out the eggs.

Raccoon invasions have increased in size and frequency in many North American marshes. At the Patuxent Wildlife Research Center in Maryland, they have become a significant threat to the duck population. Biologists at Patuxent are encouraging certain ducks to use protective nests raised on posts well above water level. The posts are covered with metal skirts that prevent the raccoons from climbing up to reach the nests.

In many parts of the country we have killed off most of the predators, such as bobcats and mountain lions, that used to eat raccoons and keep their numbers down. Because these natural predators occasionally raided farms and killed livestock, we seriously reduced their numbers. By thus removing these natural controls, we have allowed the raccoons, and other animals capable of doing much more damage, to increase in number.

On the alert for a passing marsh fish, a bobcat perches precariously on a clump of marsh grass. These primarily nocturnal cats may travel as much as twenty-five miles through the marsh each night in search of rabbits, rodents, frogs, and fishes.

A raccoon probes for eggs in a duck nest. This wary predator is on the increase because man has killed off its natural enemies. Raccoons eat almost anything, particularly the young and the eggs of other marsh animals, as well as crayfish, crabs, clams, oysters, frogs, fishes, muskrats, and birds.

Raccoons are increasing also because few people hunt them today. When I was a boy in eastern Pennsylvania, coon hunts were an exciting and profitable sport for my family and our neighbors. Often on cold, moonless nights—when raccoons were most likely to be marauding—my father and I set out across the lowlands with a .22 rifle, a kerosene lantern, a five-celled flashlight and our coonhound. Sometimes we would search in vain for an hour or more while the dog snuffled eagerly along the ground. Suddenly his excited barking would tell us he had found the scent. He would streak off into the darkness, his nose an inch from the earth, emitting low moaning barks along the way. The sounds kept receding.

Suddenly there would be a steady outpour of barks that did not diminish in volume—he had driven his quarry up a tree and was no longer on the move. We raced to join him and tilted the flashlight upward into the tree's branches. In a moment the light from two glistening emerald eyes was reflected, and we knew we had located the coon. One shot, which my father generously allowed me to make, brought him tumbling down. This was one coon who would not raid our poultry yard again. We sold the fur and made a delicious roast of the raccoon meat.

Such hunts were useful as well as exciting. We who participated in them were fulfilling the same role in nature as the bobcat or mountain lion. We were gaining a meal and incidentally preserving a beneficial balance among the living things that compete for existence in any environment.

More visitors from the uplands

Here and there along the shore of the marsh you will find holes dug in dry sand, surrounded with fragments of leathery eggshell. The skunk, another upland raider, has been at work. Skunks are fond of snapping-turtle eggs and will spend hours searching for turtle nests—and it is fortunate that they do. Snapping turtles are prolific. If there were no check on their reproduction, the turtles would become dangerously out of balance with other life in the marsh.

Deer visit the marsh at all seasons. In summer they browse on water lilies and other succulent aquatic plants. Besides eating algae, they nip off the stems of cardinal flowers, marsh marigolds, and ferns. But they come in the greatest numbers during the winter. Whole herds, unable to find enough food in the frozen uplands, crowd especially into wooded swamps where vegetation is still accessible. By spring large areas of the shrubby undergrowth may be denuded as high as the deer can reach. As a result a browse line can be found not only in our wetlands but also in upland woods where the deer population is too large for the available food supply. Now that most of the larger American predators—wolves, cougars, and Canada lynxes—have been reduced in number by hunting, man must make an effort to keep the deer population in balance. Unless we succeed in restricting their numbers, deer might become so numerous that much valuable wetland as well as other vegetation may be destroyed. On most lands, properly managed hunting can be an effective control. In national parks and wilderness areas restoration of the native predators is the best control, but even in such areas direct control of deer populations sometimes is necessary.

The snapping turtle, which feeds on any living thing it can subdue with its powerful jaws, is a predator on life in the marsh. But it is also a scavenger, patrolling the bottom for dead animal matter. The sharp-toothed rear edge of its carapace, or upper shell, distinguishes it from other turtles.

A white-tailed deer fawn drinks at a marsh pool.

The flow of energy

Your tour has shown you many typical kinds of life in a fresh-water marsh. Each plant and animal requires energy, and each animal must get it by eating something else. You examined some of the characteristic food chains, tracing the progress from the plants that store food to the animals that consume it. You also glimpsed the larger pattern of life in the marsh. Food chains cross and intertwine, composing a food web. Animals of a given species eat several kinds of plants; carnivorous animals eat several kinds of plant eaters. When one species grows especially numerous, other species are deprived of their customary food. The destruction of one sort of animal may allow others to multiply at such a disastrous rate that mass starvation results. These energy dependencies and rivalries determine the character of life in every habitat. You will find the same kind of story, although many of the elements are different, in a visit to another kind of marsh—one of the flat grassy wetlands that lie along our seacoasts.

Food in the salt-water marsh

Mussels! Have you ever seen so many in one place? They are packed into the muddy bank like sardines in a can. Try to pull them out and you will have a job on your hands —they are anchored firmly to the bank by tiny fibrous growths called *byssal threads*. As you might guess from their appearance and habitat, mussels are related to clams, oysters, and other mollusks and depend on water to bring them their food. The ones we are looking at in this bank are inactive for the moment. With their shells closed tightly, they are waiting for the ocean tide to return. When the water reaches them, they will open their shells immediately. Each mussel will put out two little siphons and begin to pump water through its body. In the course of an hour the mussel will take in at least a gallon through one siphon and expel it through the other. As the water flows through the

In salt marshes such as this one on Cape Cod, the two dominant grasses are salt-marsh cordgrass and salt-meadow cordgrass. The salt-marsh variety grows where there is maximum exposure to tidal flooding, whereas the shorter salt-meadow cordgrass grows where it is drier.

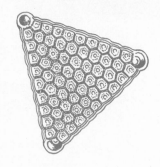

A diatom of the order Pennales is one of some 5500 species of this one-celled alga, a basic food of both fresh- and salt-water animals. The fossil remains of diatoms are used in toothpastes, scouring powders, plastics, paints, insulators, and filters.

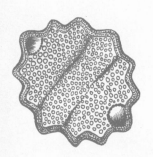

One cubic acre of water may contain many tons of diatoms in an infinite variety of shapes. They fall roughly into two recognized orders: the Centrales, which are round, triangular, or some variation on those shapes; and the Pennales, which are elongated. The Centrales apparently do not move under their own power, whereas many, but not all, of the Pennales propel themselves in slow, jerky motions, forward or backward only.

mussel's body, it must pass through tiny openings in the gills. These strain out the microscopic algae and animal life as well as the fine, partly decomposed plant and animal material—*detritus*—in the water. Although much of it is dead, this detritus is organic and still contains enough energy to support the growth of the mussel in its sedentary existence. We eat a small amount of detritus whenever we dine on mussels, clams, or oysters.

If you take a sample of water from a tidal creek and examine it under the microscope, you will find various kinds of living food as well as detritus. Diatoms are present in great abundance. These single-celled organisms surround themselves with two overlapping shells of silica, the same substance that makes up quartz. Sometimes the cells come together in large clusters or colonies; they may join together in loose chainlike aggregates or radiate outward from a common center in a star shape.

Diatoms are mostly microscopic, but some of the colonies are large enough to be visible to the naked eye. They occur either as freely floating forms collectively called *plankton* (from a Greek word meaning "wandering") or as clustering thousands clinging to submerged rocks and plants. Stones in the water often feel slick because of the presence of dense, oily diatom populations. The scouring action caused by severe tides periodically sets these colonies loose, and they become wanderers in the water, eaten eventually by animals such as shellfish and certain finfish.

The *dinoflagellate* is a common inhabitant of tidal water. This microscopic, single-celled alga has two grooves, one circling its body and the other running lengthwise. A whiplike hair, or *flagellum*, extending from the longitudinal groove propels the dinoflagellate forward, and the flagellum that circles the organism helps control its direction.

Both these yellow-brown algae—dinoflagellates and diatoms —are essential food sources. Like green plants, they possess chlorophyll, but the green color is masked by yellowish-brown pigments associated with chlorophyll. In the presence of

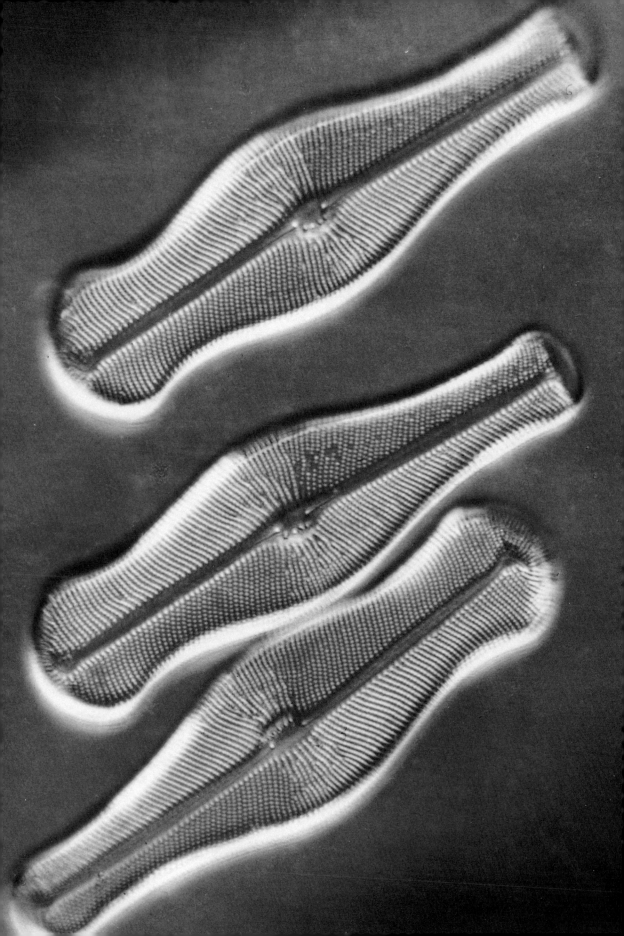

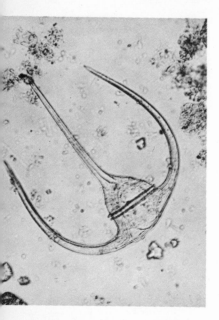

The single-celled dinoflagellate, a fundamental source of food for other life, moves through tidal waters by means of whiplike hairs located in grooves on its body. Dinoflagellates have chlorophyll in their tiny bodies and, therefore, can manufacture their own food. However, they appear to have a golden color, because the green chlorophyll is masked by another pigment.

light, these organisms are able to manufacture their own food, and they regularly store up carbohydrates and oils in their own bodies. When they are consumed by shellfish and other marine animals, they are supplying energy to creatures who cannot get it directly from the sun.

Algae in the mud

Even the black oozy mudflats contribute nourishment to the marsh dwellers. The mud algae that live there are mainly microscopic, though some of the larger green and brown colonies can be detected by an investigator with sharp eyes. The tide occasionally carries such algae away, and they end up in the digestive apparatus of a clam or mussel. The rest eventually die, and their tiny cells become part of the detritus. Larry Pomeroy, a marine biologist who studied mud algae in Georgia, found them unusually vigorous. They carry on their photosynthetic activity in all seasons. During the hot summer, they make their food when the tidal water cools the mud to the proper degree. In winter, they grow best when the tide is out and the sun warms the mud sufficiently. Pomeroy used a simple device to determine what conditions are especially favorable for the photosynthetic activity of these algae: he covered sections of the mudflats with wide-mouthed glass jars. At intervals throughout the year he measured the amounts of oxygen—a waste product of photosynthesis—present in the jars. When the oxygen content of the jars was highest, it was clear that the algae were manufacturing their food at the highest rate. Pomeroy was thus able to determine the amount of photosynthetic activity of these plants during both winter and summer.

Fiddlers and other crabs

If you look closely, you will find inhabitants much larger than algae and diatoms in tidal marshes. There are numbers of small holes around the clumps of grasses. Some resemble little chimneys barely an inch in diameter; others are a good deal wider. But you will have to be patient to catch a glimpse of the secretive creatures that have dug these holes.

A pair of stalked eyes emerges slowly from one of the small holes and peers around; then two claws, one much larger than the other, hoist themselves out of the hole; finally you see all of a male fiddler crab, so named because he moves his disproportionately big claw in a way that reminded someone of a violinist playing his instrument. Fiddlers come out of their burrows at low tide to forage in the detritus left behind by the retreating waters. Decaying vegetable matter and tiny organisms in the mud are their chief food. The fiddler uses his smaller claw for feeding by carrying little clumps of mud to his mouthparts. Fine bristles around his mouth screen out the edible particles and reject the rest.

Fiddlers rarely use their big claws for feeding, but during mating season these claws play an important role. A male fiddler will stand at the entrance to his burrow for hours, waving his claw strenuously to gain the attention of passing females. Sometimes he spends two or three days signaling before he attracts a mate.

The larger holes you noticed belong to another species, the purple marsh crab. With luck you can find one lurking at the top of its burrow until an unwary fiddler passes by. Then there is a swift burst of action, and the fiddler, no match for the larger crab, is dragged underground. Sometimes,

The red-jointed fiddler crab, largest of the fiddler species, lives on the grasses and algae of salt marshes and brackish estuaries of the Atlantic Coast as far north as Buzzards Bay in Massachusetts. The eyes of this industrious scavenger are on jointed stalks that can be raised or lowered like periscopes.

however, the purple marsh crab ends up with only the large fiddle claw of his intended victim. The fiddler sheds the claw readily and grows a new one in its place.

Both these crabs play a basic role in the economy of the marsh. They go directly to the primary sources of energy, the algae and the grasses. Eating these, they break them down into finer material that is subsequently available to the other members of the marsh community. Their fecal pellets still contain energy other animals can make use of. Their excrement joins the detritus and may eventually end up inside a mussel, clam, or oyster. The nutrients released by crabs on the marsh thus resemble the contributions of earthworms in a field or lawn.

Birds of the salt marsh

Egrets are much more common in our northern marshes than was formerly the case. You can see at a distance five or six of these tall white wading birds. Study them through your binoculars. Each is standing absolutely immobile in

Dining on a piece of cordgrass, a purple marsh crab devours the plant by grinding it down with a side-to-side action of its mouthparts. Purple marsh crabs live in salt marshes, where they construct burrows amidst the plant roots. They are primarily vegetarians, but vary their diet with an occasional fiddler crab.

the shallow water, eyes looking downward, silent and attentive. Suddenly, with lightning swiftness, the egret's head darts downward and returns; the long spearlike bill has stabbed and captured a fish or crab.

Other birds are seen frequently in northern salt marshes. The little seaside sparrow darts to and fro. Sea gulls circle overhead, calling hoarsely to one another. You hear the odd *chah-hah-hah* of the clapper rail, one of the typical species of the grassy marshes. About twelve inches high, with an inconspicuous brownish-gray body and a stubby tail, the clapper rail spends its time among the grasses and tidal creeks, feeding on clams, crabs, worms, and insects. *Amphipods,* also known as scuds, are its favorite food. (You can find these tiny hopping creatures by pulling the wet marsh grasses apart.) The scud's shrimplike little body is flattened and covered with shiny scales. It takes a large number of amphipods to nourish a clapper rail on an average day.

The clapper rail nests in the marsh just above the high-water mark, and, even though it weaves a concealing can-

Laughing gulls scavenge along coasts and over salt marshes, picking their food from garbage dumped in coastal waters, or turning up worms, crustaceans, and shellfish by treading on wet mud or sand. From time to time they carry shellfish high into the air, then drop them in order to crack open their body armor on the rocks below.

THE SKINNY RAILS

Rails are often heard but seldom seen in the dense marsh. They are secretive birds that haunt the night and twilight with their cries but remain hidden in the undergrowth. In flight, rails flap awkwardly at low level, with their long legs dangling over the marsh grasses. They swim even less gracefully. Rails have narrow, compressed bodies, an adaptation for slipping through thick moist grasses and sedges. It is this slenderness that originally evoked the phrase "thin as a rail."

Rails belong to an ancient family that dates back some seventy million years. Today they are among the most widespread of birds, living mostly in marshes and feeding on a wide range of animal and plant life. There are six different rails in North American marshes—the Virginia, clapper, yellow, black, king, and sora.

The sora, the most common rail in North America, inhabits both salt- and fresh-water marshes. Now and then it emerges from the dense marsh vegetation to forage for insects and mollusks.

The clapper rail lives in salt marshes. When attacked, it sometimes dives underwater and stays down by gripping the base of a reed until danger has passed.

The king rail (opposite) is not so secretive as most rails. It lives in fresh-water marshes and feeds mostly on seeds.

The eggs of the clapper rail, like the bird itself, are concealed in the salt marsh by protective coloration. The clapper rail survives despite large-scale predation, because it lays as many as a dozen eggs often twice a season.

opy of marsh grasses over the nest, its eggs are never safe. Coons, skunks, minks, and opossums often rob the nests and suck the eggs dry. Southern water moccasins have been known to consume both eggs and young birds. Hawks sometimes successfully assault the nest, but the female clapper rail is brave. She will spring from the nest and strike back courageously to protect her young from the assailant. Severe coastal storms sometimes flood the marshes and destroy the clapper-rail nests. In the past, the most serious predator has been man, who used to hunt the clapper rail on moonlit nights at high tide. At such times the birds rise to the top of the grasses, where they are easy targets. This pastime has lost its appeal for many sportsmen, and life is now safer for the clapper-rail populations. Hopefully, these interesting birds will remain safe so long as we maintain sufficient salt marshes for them to breed and raise their young in.

All day long, predatory birds circle the marsh. You sight an osprey, or fish hawk, high in the air above a tidal creek. This bird's keen eyesight enables it to detect even a drab-colored fish swimming well below the surface. Suddenly, with incredible speed, the osprey dives into the water and seizes the prey with both talons. After it makes a catch, it returns to its nest, carrying the prey in a plane with its body, with one talon in front of the other. Some ornithologists think that this flight posture reduces the wind resistance the osprey must overcome. Others think it is simply an efficient way for the osprey to carry a struggling fish.

With all its acuteness of vision, the osprey sometimes makes miscalculations. In a Florida estuary, one was seen to assault a fish so large that the prey dragged it under the water. It took the osprey a long time to get its claws loose and escape. Finally free, it lay gasping for ten minutes on the deck of a yacht, where it sought refuge with the little energy it had left, before it regained sufficient strength to fly away.

Ducks of many kinds come to winter in this coastal marsh, which is rich in food. Some feed on the aquatic plants and fishes in the shallows; others hunt where the water is deeper. Although many ducks find winter homes here, only a few species nest in the salt marsh. Mallards and black ducks spend the winter in harbors and around tidal creeks. When spring approaches, they mate and build their nests along the edges of the marsh.

92

Marsh hawks conceal their nests in shrubs and grasses. Because these birds sometimes return and add to the same nest year after year, the structure becomes quite bulky, resembling a platform. Both parents look after the young that hatch from the bluish-white eggs.

Big fish, little fish

Tidal creeks contain a wide variety of marine fishes. Small fishes such as mummichogs and sticklebacks live there the year round, feeding on mosquito larvae and tiny crustaceans.

Mummichogs, or killifish, cruise ditches in search of mosquito larvae. These "minnows" are a favorite bait of fishermen and are also often used for experimental purposes by biologists. The male mummichog turns a brilliant orange and blue in the breeding season.

Sticklebacks have two, four, or ten spines running along their backs, depending on the species. At breeding time the male builds a barrel-shaped nest from the leaves of water plants, which are held together by mucous threads secreted from his kidneys. He courts the female by attempting to head her toward the nest. Once the eggs are laid, the bright orange male stands guard, and in about ten days the eggs hatch. Then the male destroys the nest. The adult stickleback is an important link in the coastal food chain.

Another important source of food for coastal fish and birds is the menhaden, which feeds solely on the plankton of the coastal waters. Schools of these small fish are often attacked by great numbers of sea birds from above and by predatory fish from below.

Many salt-water fish lay their eggs in the quiet coastal bays and streams, the hatching grounds and nurseries for striped bass, bluefish, flounder, croaker, and other species that take to the coastal waters when they have reached maturity. Sea bass, or striped bass, are well known to anglers and commercial fishermen. Though native to our eastern coasts, they have been "introduced" to the Pacific Coast, where they have become of considerable importance to commercial fishermen. Here is an example of an introduction that has apparently not upset the ecological balance of a natural community. Since stripers spawn in tidal waters, they have suffered greatly from river pollution, which also threatens the bluefish. Bluefish prefer the open waters of the sea, but they too breed in estuaries near tidal marshes. The survival of these important fish provides another example of how cleaning our rivers and preserving our coastal marshes benefit man in a most direct way. Tons of this delicious fish are eaten by man each year.

With its dense concentration of such microorganisms as algae, including diatoms, and with its supply of larval forms of crabs, shrimp, clams, mussels, and oysters, coastal marsh water becomes a rich feeding ground for tremendous numbers of small fishes, which in turn feed the larger fishes of deeper waters.

The mating of the stickleback is a complex ritual, a rhythm of motion and color. In spring, in the shallow waters of tidal creeks or in fresh water, the male stakes out his territory, beats off his rivals, and then builds a nest of weeds, coating it with a sticky secretion from his kidneys. When he is finished, he changes color from his customary gray to pink, then to a combination of bright orange and blue-white. He seeks out a mate and (1) courts her by zigzagging until she swims toward him, head up. Her abdomen is filled with eggs. He leads her to the nest and (2) thrusts his snout into its tunneled interior. She swims into the nest, and (3) he prods her tail, causing her to deposit the eggs. From the courtship to this point takes no more than a minute. After the female swims off, the male enters the nest to fertilize the eggs and then withdraws and searches for another female. He may run as many as four or five females through the nest. Then the male fans water over the eggs (4) and thus enriches their oxygen supply, which aids hatching. When the young are born, the male watches after them for a few days, bringing stragglers back to the nest in his mouth. Finally the young join others and set out on their own.

The living and the dead

It is easy to overlook one important part of the food web in both salt- and fresh-water marshes: the world of the *decomposers*, the fungi and bacteria that live by breaking down dead plants and animals into simpler forms. Without these decomposers, the balanced, continuing life of the marshes would be impossible.

The marsh is wet and yielding underfoot. In the deep accumulation of dead plants that constitutes part of its substance, an all-important activity is going on. We cannot see this activity, but we can observe one of its results. Bore a hole into the soft tidal marsh and hold a match to it. The match soon goes out, but a pale flame may continue to burn and flicker for half a minute: marsh gas, or *methane*, feeds the flame. Billions of microscopic fungi are at work beneath the surface, reducing the plant tissues to simpler materials and releasing methane as a chemical by-product.

The fungi and bacteria that produce this gas belong to the world of the decomposers. Like all living things, they must have energy to grow and continue their vital processes. Lacking chlorophyll, they cannot procure their energy from the sun as green plants do, and so they obtain it instead by breaking down plant and animal tissues into less complicated substances and using the chemical energy stored in them. Decomposers are constantly at work in the marshlands—and in every other habitat—converting dead organic tissue into substances they can assimilate.

Organisms that feed on living things are called *parasites*. They break down part of the host's tissue and divert the energy it contains to their own uses. Sometimes their depredations are minor. A flea, for instance, is a parasite that causes a dog or a human some discomfort but does no serious harm unless it carries a disease. Other parasites kill the organisms they invade. When water mold attaches itself to the gills of a live fish, it forms masses of *mycelia*, threadlike vegetative parts, which prevent the entry of air and suffocate the fish.

Once the fish is dead, another kind of water mold takes over. This type, which attacks only dead organisms, is called a *saprophyte*, from two Greek words meaning "plant growing on the dead." Millions of saprophytic spores are present in every handful of ooze you scoop up from the marsh. The bacteria it contains are especially numerous, and so long as

The jellylike *Tremella* develops on decaying wood in wooded swamps and moist forests. This decomposer, along with countless others both visible and microscopic, is a vital link in the cycle of energy.

they find nourishment they multiply constantly. They do this, as do other cells, simply by dividing in two over and over again. The reproduction rate of bacteria is hard to imagine; it has been estimated that one cell reproducing uninterruptedly under ideal conditions for twenty-four hours could develop into seventeen million individuals.

Molds and fungi are of many sizes and varieties. The mushrooms growing on dead logs in the swamp forest and the tiny bacteria that can be seen only with a high-power microscope both live by breaking down other organisms and annexing a portion of the energy they contain. This work of decomposers is vital to all life. They restore materials to soil, air, and water, making them available to plants again. If it were not for the decomposers, all the materials necessary to support life would long ago have been "locked up" beyond recovery.

A bog mushroom pokes up through the sphagnum mat. Minute threads, or hyphae, work their way downward into dead matter, penetrating microscopic intercellular spaces and freeing nitrogen compounds and other substances upon which the fungus thrives.

Animal decomposers

Now, with an understanding of the last link in the chain—the role of the decomposers—you are ready to draw a diagram showing how energy, coming initially from the sun, flows throughout any habitat, in this case a marsh. First you should complete your understanding of decomposers by returning for a moment to a fresh-water marsh for a quick look at some animals at work there. If you scoop up samples of ooze from a cattail marsh or a bog, you are likely to find several kinds of worms struggling to escape. Their bodies are segmented and bristled like those of earthworms, to which they are related. These bristle worms plow the marsh bottom as earthworms plow lawns and fields. They pass surprising quantities of dead plant material through their bodies every day, retaining some for food but discarding the rest.

Some bristle worms are so transparent that you can see through them; others, such as *Tubifex*, are quite red. They build little tubes or tunnels to live in, and all you usually see of them is their waving red tails protruding from the mud. The brilliant red bloodworm *Chironomus* also builds a little tube on the surface of the mud. The bloodworm is actually the immature, or larval, stage of a midge, a type of fly. Like some of the fungi we have discussed, the midge larva can live where oxygen is scarce, and so you often find it living in the deep marsh ooze. Other kinds of bloodworms live in tidal pools, where they feed on algae and detritus and play an important role in the vital process of decomposition.

The endless cycle

As you have seen, each type of decomposer takes what it needs from the organic matter in its environment and leaves the decaying remnant for other varieties to decompose further. Materials essential to the food-making activities of plants are released at each step.

The coral fungus is named for its large, colorful, corallike intertwining branches. It grows on fallen trees and in rich humus in swamp forests and wet woodlands. Rodents often devour the coral fungus within hours after its emergence.

The flealike amphipods scavenge for carrion; midge larvae feed on dead and living plant tissues; and *Tubifex* worms living head down in mud recirculate detritus by expelling wastes from their tail ends. Thus, these and other scavengers capture the energy remaining in the dead material in this endless cycle.

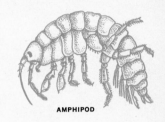

AMPHIPOD

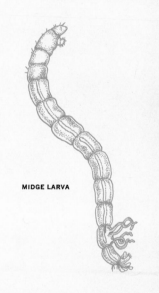

MIDGE LARVA

TUBIFEX WORMS

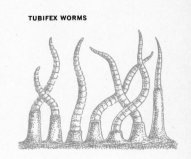

99

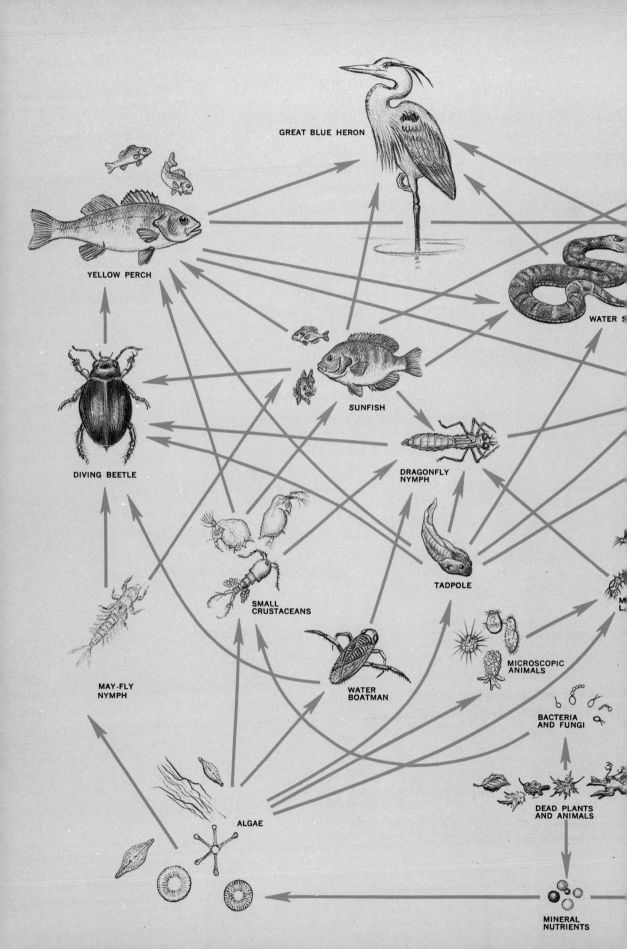

GREAT BLUE HERON

YELLOW PERCH

WATER S

DIVING BEETLE

SUNFISH

DRAGONFLY
NYMPH

TADPOLE

SMALL
CRUSTACEANS

MICROSCOPIC
ANIMALS

MAY-FLY
NYMPH

WATER
BOATMAN

BACTERIA
AND FUNGI

ALGAE

DEAD PLANTS
AND ANIMALS

MINERAL
NUTRIENTS

SNAPPING TURTLE

BULLFROG

MALLARD

TUBIFEX
WORMS

MINK

MUSKRAT

LARGE PLANTS

THE LINKS OF ENERGY

This food-web diagram shows the pathways of energy. Each arrow leads from prey to predator. Although an oversimplification of reality, the diagram is useful for depicting the flow of energy and showing how change would alter that flow. In effect, the web is made up of interlocking food chains. If, for example, the numbers of May-fly nymphs and small crustaceans were reduced, a corresponding reduction in sunfish, their consumer, would probably take place. Fewer sunfish would mean less food for the herons, thus fewer herons. Fewer herons would probably lead to a superabundance of water snakes, normally the herons' prey. In reality, nature is more dynamic and capable of adjustments. Sunfish, for instance, might survive the depletion of May-fly nymphs and small crustaceans by relying on other sources of food. At the far right of the web is man, who, unlike other creatures, can destroy or endanger life for purposes other than food — to collect the fur of muskrats and minks, to gather and sell fashionable plumes, for sport — or for no purpose, such as happens through carelessness. Man is the unique predator of the web — and potentially the most dangerous.

These materials are really fertilizers that permit the green plants to continue to live and grow. Back they will go into cattails, bulrushes, marsh grasses, and algae, helping these plants to manufacture the food that will nourish muskrats and tadpoles and, later, minks and snakes. The energy this food contains is used up by those who eat it, but none of the matter itself is lost. Once it has been broken down, it is ready to be used again. So long as the sun shines, life will continue.

Most forms of life can live only because of the presence of other forms. Many members of a species must die, to feed another species. But the disappearance of any one type of life could seriously modify the pattern of existence in the marsh. In the long run a balanced natural community is maintained here, as in all natural environments. Man, by intervening and killing off a species, has made tragic blunders and done much harm. But little by little he is coming to understand the order and harmony of the natural world and to see the importance of the food web of which he is a part.

The coastal and fresh-water marshes bring together many kinds of life, and some are rivals for the same type of food. Many become food for larger animals. All the plants and animals mentioned have a place in this complicated pattern, this food web. Ecologists, the scientists who study these interactions and dependencies, attempt to interpret the total marsh environment—water, soil nutrients, and air—as well as the plants and animals in the ecosystem. One eminent ecologist has observed that this interacting complex, the ecosystem, is not merely more complex than we think, but rather more complex than we *can* think.

In the pattern of nature, life is served by death. By killing a mallard, this red fox, like other marsh marauders is helping to maintain the delicate and vital balance among living things.

The Ways
of Wetland Life

Plants and animals living in marshes, swamps, and bogs
have evolved a fascinating array of adaptations that fit them
for the unique conditions of life in their part water, part land
environment. For example, coastal marshes are washed by
ocean tides, and salt-marsh cordgrass, reeds, mangroves,
and other plants have become adapted to this regular bath
of salty water. Farther inland, in fresh-water marshes, dragonflies have developed in two worlds: they begin their lives in
the water, then emerge as adults to soar over the marsh in
search of prey. Duckweeds float unrooted in the open water;
closer to shore rooted water lilies grow their large floating
leaves; still closer to shore, other rooted plants—pickerelweed, arrowhead, and cattails—rise above the water surface.
Each kind of plant and animal lives only where conditions
are within its range of tolerance.

Whatever the special adaptations of plants and animals—
whether to the tidal rhythm that regulates the fiddler
crab's activity or to the salinity that dictates where cattails can and cannot grow, or whether they are the actions
and ultimate death of a beaver if the pond in which it
lives freezes to the bottom—they are part of a pattern of

Were it not for the large airspaces in the stems of pondweeds and many other aquatic plants, their leaves could not stay afloat, because the weight of the stems would drag them under. The airspaces make the stems light and buoyant no matter how many feet in length they may grow.

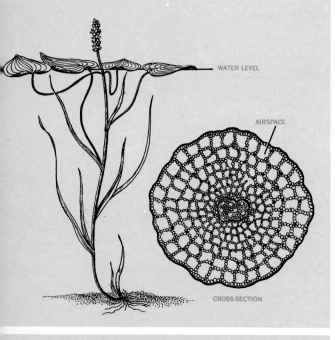

WATER LEVEL

AIRSPACE

CROSS-SECTION

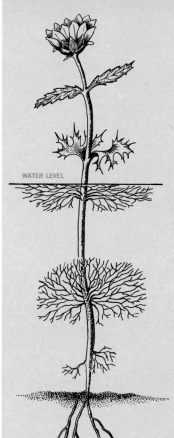

WATER LEVEL

Water marigold has two kinds of leaves; one kind extracts gases from the a the other from water. Leaves that project above the surface of the water take in and release gases through their stomates, b the finely divided under-water leaves absorb and give off oxygen and carbo dioxide directly through their cell walls.

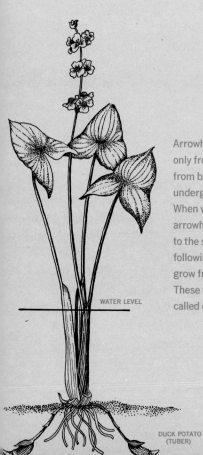

Arrowhead reproduces not only from seeds, but also from bulblike tubers on underground stems. When winter comes, the arrowhead plant dies down to the stocks, and in the following spring new plants grow from the tubers. These tubers are popularly called duck potatoes.

WATER LEVEL

DUCK POTATO (TUBER)

The floating leaves of spatterdock have stomates on their dry, upper surfaces. On most plants these pores, through which carbon dioxide and oxygen pass in and out, are on the undersides of leaves. The large airspaces in the leaves help them stay afloat.

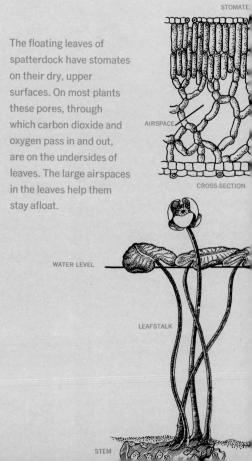

STOMATE

AIRSPACE

CROSS-SECTION

WATER LEVEL

LEAFSTALK

STEM

The female flowers of wild celery rise to the surface of the water on long stems, then open. The male flowers develop underwater, break loose, and float to the surface where they open and drift about until they contact a female flower. After pollination the stem of the female flower coils downward, and the seeds mature underwater.

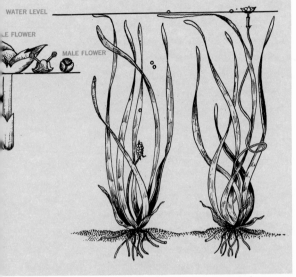

WATER LEVEL

LE FLOWER

MALE FLOWER

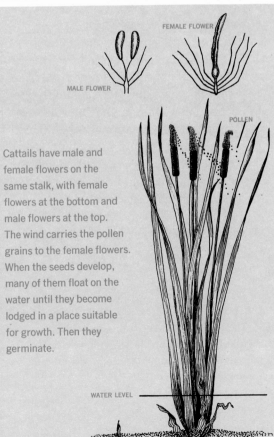

FEMALE FLOWER

MALE FLOWER

POLLEN

Cattails have male and female flowers on the same stalk, with female flowers at the bottom and male flowers at the top. The wind carries the pollen grains to the female flowers. When the seeds develop, many of them float on the water until they become lodged in a place suitable for growth. Then they germinate.

WATER LEVEL

SURVIVAL IN A WORLD OF WATER

Aquatic plants, or hydrophytes, have evolved a fascinating array of adaptations that allow them to live in their unique habitat. The devices by which they cope with a superabundance of water and, sometimes, a complete lack of atmospheric oxygen are among the most interesting in the natural world. One adaptation common to nearly all aquatic plants is their spongy tissues. Large airspaces in stems and leaves help keep many of the aquatic plants afloat. Moreover, they help to distribute the carbon dioxide and oxygen necessary for the plants' metabolic processes. Often the underwater parts of hydrophytes are markedly different from those abovewater. In some, specialized leaves, only a few cells thick, absorb gases from the water directly through their cell walls. Whatever the adaptation, whether the plant is a floater, an emergent, or a submergent, the plants in marshes, swamps, and bogs are so perfectly adjusted to life in water that they not only survive but flourish there.

behavior that reflects an adaptation to the conditions of the wetlands.

Plants of the cattail marsh

The green plants in any wetland occupy many and unusual places in the habitat and vary greatly in structure and size. Some of the most important green plants, as you have already discovered, are so small that they cannot be seen without a microscope. Some are so large and grow so densely that you can hardly push your way through them in a canoe. Some grow under the water, some float on top, and some stand high over the water. Yet all these plants have the same basic requirements—sunlight, water, carbon dioxide, oxygen, and minerals—though they may meet these needs in many different ways. They multiply in a surprising variety of ways, too.

Rooted or floating, submerged or emergent, with broad leaves on the surface or ribbony underwater leaves (or both), aquatic plants make a fundamental difference to the other living things with which they share the water. Those that are rooted serve as bases to which protozoans and algae can attach themselves. They give concealment to crustaceans, insects, and fishes, enabling many species to elude their enemies for a time and populate the waters.

On your tour of the marsh you have seen how rooted plants can in time cover the marsh and impede your passage through its waters. Eventually they may fill it completely and destroy its essential character, but in the interim they make indispensable contributions to its animal inhabitants. They provide support, shelter, oxygen, and nourishing food. Even when the plants die and are decomposed by decay organisms, these nutrients are again made available to the minute drifting organisms as well as larger rooted and free-floating plants.

On the other hand, the continuity of plant life is depend-

Cattails (*left*) are among the most easily recognized of all emergent plants. Only their roots and the lower portions of their stems are underwater. Water lilies (*right*) are floaters. Their buoyant leaves are connected by slender leafstalks to thick stems buried in the bottom sediment. The submergents, another group of wetland plants, live completely underwater.

ent upon, and often modified by, the presence of animals. Ducks and other birds disperse great quantities of seeds from one wetland to another. Because plant eaters often have different preferences, competition between plants is reduced and individual species are prevented from crowding out others. For example, ducks especially like pondweeds and duck potatoes, whereas the muskrat prefers a cattail stand as a food source. But animals can do damage, too, as when deer "overbrowse" in winter and destroy much of the shrubby undergrowth, or when plant eaters destroy the plant life of a marsh after the meat eaters that feed upon them are killed off.

Pollination

Animals, especially insects, perform an essential service in the reproduction of many seed plants by carrying pollen. Thousands of plant species with heavy or concealed pollen have no other means of effecting this transfer. Without the intervention of insects, some might die off in a single generation.

Nearly all flowering plants reproduce sexually. In certain species, both male and female sex cells are present in the same flower or adjacent flowers, and the plant is able to pollinate itself. But more often the *stamens,* or male parts, and the *pistils,* or female parts, mature at different times. When this happens, external agents are needed to convey the pollen from one flower to another. A similar problem occurs when male and female flowers grow in separate clusters on the same or on different plants.

Three agents are principally responsible for pollination: wind, water, and insects. For at least 135 million years, since their origin in the Cretaceous period, flowering plants have depended on the casual action of winds and water and on the foraging of bees, birds, and butterflies to convey their pollen to the female parts. Certain plants are remarkably adapted to facilitate these blind operations.

The cattails, sedges, grasses, and rushes of the marsh and many of the shrubs and trees along its banks are wind-pollinated. In early spring, the branches of alders have not yet sprouted leaves, but the male flowers can be seen drooping from the twigs—bright two-inch-long festoons called *catkins.* Tap one and you will see the fine powdery yellow

Wind pollinates the pussy willow, scattering millions of pollen grains throughout the marsh. These are male catkins gleaming brightly with the pollen they contain. The female catkin is a less resplendent gray-green.

pollen ride off on an imperceptible current of air. Inside the pollen grain the male sex cell is being carried safely. If it alights on a female alder catkin nearby, fertilization may occur, and seeds will develop. Though the odds against pollen reaching a female flower are incredibly high, there may be millions of catkins in the marsh, and literally billions of pollen grains make the journey every year. Many reach their destination, and the alder population is maintained.

Emergents and floaters with bright conspicuous flowers, such as water lilies and pickerelweed, depend mainly on insect pollination. In the marsh as everywhere else, distinctive scents and vivid colors are adaptations that attract insects and other pollinators. The nectar produced by the blossoms contains nourishing sugar that attracts innumerable bees, butterflies, and moths. Some of the pollen these insects carry off on their legs will adhere to the sticky pistils of the next plant they visit.

A bee, drawn to the tiny but colorful flowers of the water smartweed, brings pollen from other smartweed plants. Although floaters that are insect-pollinated have flower stalks abovewater where insects can reach them, some have stalks that wither after pollination, their fruits developing underwater.

Totally submerged plants either pollinate themselves or depend on water to carry their pollen to female flowers. The pollen grains of some submergents are specially adapted to underwater movement. Released in great quantity, they are borne at random by the current, and some of them eventually encounter a female flower of the same species.

Other submergents, among them water celery, waterweed, and eelgrass, have a more complicated system. The water celery that often chokes the channels of marsh watercourses has male flowers that break off underwater and then float to the surface, where the pollen grains sail around in little "boats." Meanwhile the female flower sends up a flower stalk abovewater. In time a pollen boat collides with a female flower, and if some of the pollen adheres to the pistil, pollination occurs. After pollination, the female flower stalk is withdrawn below the water, where the fruit develops.

From seed to seedling

Most marsh plants have acquired varying adaptations that favor their dispersal. The seeds of water lilies, bur reed, and arrowhead are adapted to float on water currents; sometimes they establish themselves two or three miles from

The seeds of the red mangrove of southern Florida are protected from the lethal salts of the swamp by developing into seedlings on a tree. When the seedling is about a foot long, it drops from the parent plant and floats until the sharp root tip anchors. From this initial grip the young mangrove shoots out stiltlike prop roots.

their source. The airspaces they contain make them buoyant, and they can travel for months until they come to rest on a mudflat. Other seeds, such as those of cattails, willows, and *Phragmites,* may ride the air currents for several miles before alighting. Birds also aid in the dispersal of seeds; Charles Darwin found as many as five hundred different seeds adhering to the feet of birds. Many seeds that are eaten by mammals and birds pass through the digestive tracts of these animals unharmed and are often distributed over great distances. This is why plants that do not have floating seeds are able to become established on islands far from the mainland.

Many types of seeds not only can travel long distances but also can lie dormant for many years. Their tough outer covering protects them from harsh changes in the environment. When, by chance, a seed finds itself in favorable conditions, life stirs inside it, structural modifications take place, and it puts out roots and a shoot and becomes a seedling. But the environmental hazards most seedlings face are so great that many are lost. Both seeds and seedlings are eaten in enormous quantities by insects, birds, and rodents. The young seedlings are often destroyed by floods, abrupt changes in weather, or heavy shade. Only the great number in which most seeds are produced assures the success of a few.

Amphibious insects

Brilliantly colored insects dart through the air above the cattail marsh, their wings glittering in the sunlight. In their swift, purposeful flight they are searching for other insects, chiefly mosquitoes and flies, to feed on. Their appetites are so voracious that they sometimes eat one another. Dragonflies and damsel flies are the two most conspicuous kinds of insects seen in the marsh.

When these insects alight on pickerelweed or arrowhead, you have more time to study their finely netted wings and slender bodies. An expert can identify each species by its color pattern and markings. Blues, reds, and browns predominate. Those individuals that hold out their wings horizontally when resting are true dragonflies. The others, which fold their wings vertically above their bodies when they rest, are damsel flies. Of the two thousand different species of this order, three hundred have been observed in North America.

Damsel flies and dragonflies have similar life histories. They hatch and spend their immature phase in the water, never flying until they reach adulthood. You can find some

114

of these immature forms, called *nymphs,* crawling on the muddy bottom. The damsel-fly nymph has three little gills that look like flags fluttering at the rear of its body. In its aquatic state, it feeds on a variety of animal food. The dragonfly nymph, too, is a carnivore from the moment it hatches. Folded under its head is a long specialized mouthpart that the nymph can shoot out with lightning speed to seize passing insects and other animals, including small fish.

The dragonfly nymph also has gills that can take the oxygen directly from the water, but unlike damsel-fly gills these are internal. The dragonfly nymph takes in water at the rear of its body and pumps it through a network of tubes in the gills, where the oxygen is extracted and carbon dioxide is discharged. The pumping apparatus is surprisingly forceful. Examined in a collecting pan in the laboratory, the dragonfly nymph will lie quietly for a long time, as though dead. Suddenly like a jet it drives itself forward with astonishing speed by expelling the water from the gill chambers in the rear of its abdomen.

The dragonfly holds its wings spread out when resting. One of the fastest of insects, it has been clocked at flight speeds up to a mile a minute. In midair, the dragonfly scoops its prey into basketlike legs. Its bulging compound eye, set in a swiveling head, is composed of thirty thousand separate eyes, each with its own lens and retinal mechanism.

THE LIFE CYCLE
OF DAMSEL FLIES

When damsel flies are about
to mate, the male transfers
sperm cells from his ninth
abdominal segment to a
bladderlike pouch on his
second abdominal segment.
Then the female, often gripped
at the back of the head by a
special clasping device at the
tip of the male's tail, thrusts her
body forward and upward from
below, inserting the tip of her
tail into the sperm-containing
pouch. After mating, the female
deposits the eggs through her
ovipositor in submerged
vegetation or in sandy shallows.
She often will dive as much as
a foot underwater to do this. . . .

2

... The time required for the
eggs to hatch depends upon
the kind of damsel fly and such
environmental conditions as
water temperature. Nymphs,
the second stage in the life
cycle, may emerge after a few
days or as late as eight or nine
months later. When they do,
they become voracious feeders,
snapping up the larvae of gnats,
flies, mosquitoes, and other
water-dwelling insects. Finally,
when the nymph is fully grown,
its skin splits and the adult
damsel fly begins to emerge (1).
As blood plasma is pumped into
its wings (2), they begin to
harden. When the wings reach
full size, the abdomen begins
to expand (3) and continues to
do so until it reaches its normal
proportions. The filling out and
drying of the wings may take
only a few hours, or it may
take several days.

3

Whirligig beetles are often present in large numbers in fresh-water marsh pools when the weather is warm. In winter they are active only on warm days. Their compound eyes are divided into two parts: one for seeing above the water surface and one for seeing below it — a handy adaptation for catching insects both under the water and on it.

Nymphs of both groups feed and grow in the bottom mud until they are ready to molt and become adults. Then they crawl up the stem of an emergent plant and shed the skin that has encased them. After their wings unfold and harden, they fly off in search of food or mates.

Some damsel-fly species actually go underwater to lay eggs. They fold their wings around their bodies to capture a sufficient supply of air and walk down plant stems to lay their eggs below the surface. Scientists have observed female damsel flies spending as much as twenty minutes laying eggs two feet underwater.

If you look attentively, you can find an odd little parcel of tiny leaf fragments and sticks moving deliberately along the bottom of the water in the wooded swamp; you can see legs protruding from the tube. This is a caddis-fly larva, hiding inside the camouflage it makes out of dead leaves and twigs. It feeds on small water plants and animals during its aquatic stage. Wherever it goes, it drags its case along.

The caddis fly spends all but a tiny fraction of its life in the immature, aquatic stage. When a caddis-fly larva sheds its skin and molts into the flying form, it looks like a drab little moth. If you examine it closely, you will see that its wings are covered with tiny hairs, instead of scales like those of moths and butterflies.

Other marsh insects

Wherever there is standing water in the cattail marsh, you will find many of the insects that regularly live in ponds. The water strider is one. From the onset of spring until late autumn you can see these long-legged bugs skating and leaping about on top of the water. Water striders can do this without sinking because of the surface film on the water. Liquids are always especially dense at the surface. Molecular attraction comes only from below there, and the outermost layer holds together like a tough, flexible skin. The striders move about vigorously and safely on this layer. They are carnivores, feeding on backswimmers and emerging midges. One species can leap several inches into the air to catch mosquitoes in flight.

Whirligigs are present, too. They get their name from their habit of spinning rapidly about on the water surface, creating the appearance of miniature speedboats. Catch one —if you can—and notice its apple-seed odor. Possibly this scent protects the whirligig from enemies. Whirligigs can fly, but they must first climb out of the water on a plant stem.

The water boatman, usually less than half an inch long, is one of the smaller residents of the marsh. It swims underwater in a quick darting manner, using its hind legs like oars. This insect spends a good deal of time feeding on decayed debris, diatoms, and filamentous algae. It literally sucks the contents right out of the filaments of the green alga *Spirogyra*. Since the water boatman has no gills, it envelops almost its whole body in a bubble of air and carries this bubble along with it when it dives to the bottom of a marsh pool in search of food.

These are only a few of the insects you can easily find in marshes and swamps. Some, such as dragonflies and damsel flies, are amphibious, spending part of their lives in water and part on dry land. Others never leave the water. In getting their much-needed oxygen, they depend upon gills, air bubbles, or even breathing tubes thrust to the surface as in the case of mosquito larvae. Some insects of the marshlands are exclusively carnivorous; some eat only vegetable matter; others are scavengers living on dead animals and plants. Whatever their diet, these insects of the marshes and wooded swamps constitute an important link in the food chain and show an infinite variety of adaptations to an aquatic environment.

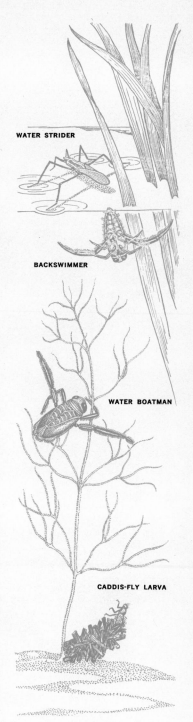

WATER STRIDER

BACKSWIMMER

WATER BOATMAN

CADDIS-FLY LARVA

Life in and out of water—the amphibians

Among the most interesting animals of the fresh-water marsh are the *amphibians*, creatures that usually spend part of their lives in the water and the remainder on land. Amphibians are descended from fish—they were the first vertebrates to adapt successfully to life out of the water—and even today most of them still have gills at some time in their life cycle. There are a few amphibians, such as the mud puppies, that spend their entire lives in the water, and others, such as the red-backed salamanders, that spend their entire lives on land. They get their oxygen through their gills, lungs, moist skin, or a combination of these ways.

There are only three kinds of amphibians in the United States: salamanders, frogs, and toads. Most of them are particularly beautiful little creatures, with colorful markings that make them exciting finds on your visit to a marsh. But they are usually shy and often nocturnal in their habits, and you must have sharp eyes to spot them. Even the huge bullfrog will startle you more often than not by jumping away from your feet as you make your way around the edges of the water.

Like most amphibians, frogs are especially active at night, when the humidity is high and their skins are not in danger of drying out. Then the chorus of spring peepers becomes nearly deafening. When you flash a light across the marsh, you must look sharply to find these tiny singers. Some will be seated under clumps of grass, and others will be riding the beds of floating algae. The bubble, or air sac, on each male frog's throat swells and subsides like a tiny balloon. For millions of years their din has risen from the marshes every spring night, summoning females to the breeding grounds where a new generation will be created.

The pretty spotted salamander often makes wooded swamp pools its breeding place. When spring comes, it migrates to the nearest body of water, instinctively traveling by night when the sun's scorching rays cannot dry out its skin. The males are first to arrive, and the females appear a few nights later. Courtships are usually brief. Prospective mates

The spotted salamander lives in moist wetlands and woodlands for all but about one month a year, when it enters the water to breed. In captivity, this amphibian has been known to live more than twenty years.

An American toad trills his mating call across the springtime marsh. Toads, unlike frogs, have warts on their bodies. The difference in length between the toad's front and hind legs is not nearly so marked as it is in frogs.

SPOTTED NEWT

RED EFT

LIFE CYCLE OF THE SPOTTED NEWT

The spotted-newt egg hatches into a gilled larva about two weeks after it is laid. When the aquatic larva is about three months old, it undergoes a marked metamorphosis. It loses its gills and develops lungs in their place, and its skin color gradually changes from drab olive to bright red or red orange. The red eft, as it is called in this stage, leaves its original habitat in the water and moves out onto dry land, hibernating during the winter months. It is most active on damp or rainy days, when it ventures forth to search for insects and other small animals. The final transformation to an adult spotted newt occurs two or three years later in the spring or early summer. Once again the salamander enters the water and changes color from red to drab olive dotted with black. Only the rows of scarlet spots from the eft stage remain.

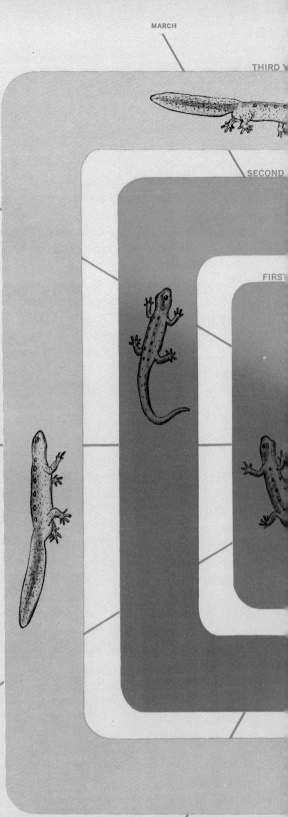

MARCH

THIRD Y

FEBRUARY

SECOND

FIRST

JANUARY

DECEMBER

NOVEMBER

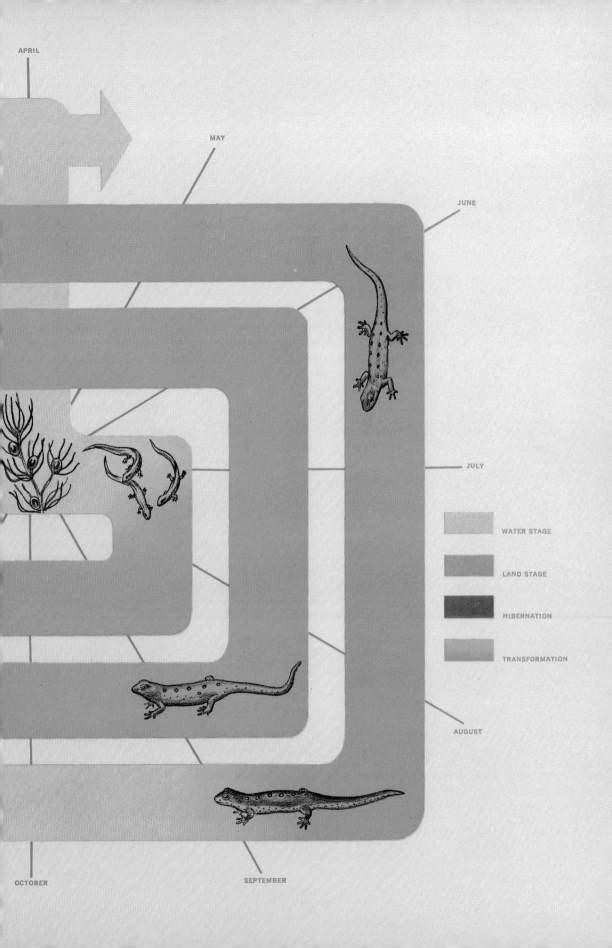

APRIL

MAY

JUNE

JULY

AUGUST

SEPTEMBER

OCTOBER

WATER STAGE

LAND STAGE

HIBERNATION

TRANSFORMATION

GEARED TO LIFE IN
AND OUT OF WATER

Though not much more than an inch long, the spring peeper — in full voice, with its vocal pouch inflated — has a shrill, piping mating call that can be heard for surprisingly long distances.

To ensure the survival of the species, frogs must lay enormous amounts of eggs. The wood frog (*below*) lays from two to three thousand eggs; others, such as the bullfrog, lay as many as twenty thousand. Each spring wood frogs migrate to the water to send out their mating calls, which sound like the quacks of a duck. The female deposits her eggs as the male clasps her tightly behind the forearms and releases the fertilizing sperm cells over the eggs. In two weeks or so, depending on the weather, the eggs, like all frog and toad eggs, will hatch into tadpoles, the larval stage. Until they mature into adults, these little swimmers will live underwater, breathing through feathery gills concealed under a little fold of skin. They will feed on the algal coatings on the leaves of submergents. Relatively few will survive the predations of fish, dragonfly nymphs, herons, snakes, and the multitude of meat eaters in the marsh.

The leopard, or meadow, frog is sometimes confused with the pickerel frog. The latter has squarish spots in two rows rather than the scattered round ones of the leopard frog.

Thousands of ducks and geese
rise from the water at Sacramento
National Wildlife Refuge in
California, a stopover on the
Pacific flyway. Awesome
congregations of waterfowl,
once a common sight in
American wetlands, are now
seldom seen except in
protected areas.

glide back and forth over one another in the shallow water.
The male occasionally nudges the female, then descends to
the bottom, discharges a jellylike capsule of sperm, and
attaches it to a stem or leaf. The female descends and inserts
the capsule into an opening in her body. Steadying herself
against a twig, she begins to expel eggs, which encounter the
sperm on the way out of her body. The fertilized eggs, surrounded with a double layer of transparent jelly, become
much larger as they absorb water. Clumps containing as
many as two hundred eggs cling to twigs. The adult salamanders depart separately and perhaps never see their young.

The eggs of spotted salamanders hatch within two to four
weeks. The young larvae resemble frog tadpoles, though you
can identify them by their three pairs of fluttering feathery
gills. After about three months the three-inch-long larvae
transform into adults, which can grow to a length of eight
inches. These harmless creatures are generally solitary and
usually nocturnal.

A salamander is not, of course, a lizard, as the ancients believed, and you can easily tell the difference between them.
A lizard is a reptile, and its skin is dry and scaly like a
snake's, but a salamander has smooth, moist skin. A lizard
usually has five toes on its front feet, a salamander only four.
Size, however, is not a good way of telling them apart, for,
if you think a lizard is large and a salamander is small, consider the giant salamander of Japan, which may grow to a
length of four feet.

126

Birds of the wetlands

Long before mankind appeared on earth, wild fowl were
following regular routes north and south over the wetlands
scattered throughout the American landscape. The first settlers marveled at their prodigious numbers. Though marsh
and shore birds became a staple in the Colonial diet, hunting did not at first reduce their quantity as seriously as the
conversion of marshes to farmland. The steady shrinking of
waterfowl and marsh-bird feeding grounds and nesting areas
resulted in great reductions in their numbers. Many flocks
were compelled to find new flight routes and new places to
feed and nest. As later settlers moved westward, they were
able to see how the birds flourished in an unmodified environment. Travelers across the prairies as late as the nineteenth century could not find words to describe the hundreds of thousands of ducks and geese in flight and the
thunder of their wings as their masses darkened the sky.

As the population of the United States grew, more and
more waterfowl were shot, especially for market sale. Some
birds, such as the Eskimo curlew, became virtually extinct.
Added to hunting pressure, the drought of the dust-bowl
years reduced duck and geese populations even further.

Even so, ducks and geese are today among the most
numerous of our migratory birds. But you should remember
that ducks and geese are very gregarious birds that congregate in large groups. Although they seem to be extremely

Despite their abundance not
all waterfowl refuges are
maintained by the Federal
government. Many privately
owned lands — ranging in size
from small backyards to
many-acred research centers —
provide food and sanctuary for
thousands of waterfowl each
year. One of these sanctuaries
is the Lockhart Gaddy Refuge,
created by a North Carolina
duck hunter who turned his
six live decoys loose on a
small pond after their use was
prohibited in 1934. Within four
years the original six birds had
been been joined by hundreds of
wild geese. By 1953 the man
had to increase the size of
Gaddy's Pond to four acres, to
support thousands of ducks,
at least twelve thousand Canada
geese (following page), and, at
times, small groups of blue,
snow, and white-fronted geese.

127

abundant, their numbers are not nearly so great as those of birds such as starlings, warblers, and blackbirds. There are about fifty-five species in the family Anatidae (including ducks, geese, and swans) living at least part of the time in continental North America. They are all adapted to watery surroundings and have common characteristics: webbed feet; short legs and tails; wide, flat bills; relatively long necks. Beyond this the similarity ceases. Among species and sub-species there is an incredible range in size, coloration, calls, and habits. Most teals weigh less than a pound; trumpeter swans sometimes weigh forty pounds. The widgeon's diet is almost entirely vegetable; the scoter rarely eats any-thing but shellfish. Each year the blue-winged teal travels from northern Canada to as far south as central Chile; the Florida mottled duck seldom leaves Florida.

The dabbling ducks

In considering how birds get along in the wetlands, you can do no better than to start with the ducks. The wood duck, found in both the eastern and western parts of the

Among the fastest, strongest, and most agile fliers, pintails are notable for their long migration, which regularly takes them from the North in early fall as far away as the Hawaiian Islands or farther. In flight both males and females are easily recognized by their long pointed tails.

Young wood ducks, or woodies, peer out from their unique nest in a tree cavity. Twenty-four to thirty-six hours after they hatch, their mother calls them to leave the nest. The dauntless ducklings step out into space and may plummet as much as fifty feet to the ground below.

United States, is an especially instructive example. On the one hand it likes to nest well off the ground, like many an upland bird, and on the other hand it is particularly fond of foraging in wooded ponds, swamps, and marshes.

Wood ducks nest in hollow trees or tree limbs, three to fifty feet off the ground. The young are provided with sharp claws that help them climb from the bottom of the nesting hole to the entrance, but as soon as they flutter, or simply fall, to the ground, they take to the water just like any other ducklings. Like most dabbling ducks, the wood duck searches the ground as well as the water for food. Its legs are set well forward on its body, and it gets about easily on land, where it finds seeds and nuts. In the water it eats duckweed, tubers, and occasional insects. If it is suddenly startled as it paddles about in coves surrounded by trees or in and out of cattails, it takes off directly from the water, jumping into the air in a small space.

On your visits to the wetlands, then, watch particularly for these dabbling ducks. Among them is the mallard, probably the most familiar and common of our wild ducks. The

Of all the wild duck species in
the world, the hardy mallards
are the most plentiful. They
are fast-flying and can launch
themselves with a burst of
speed almost vertically off the
surface of the water into the
air. Typical of surface-feeding
ducks, they prefer shallow water
where they dabble, tail up
and head down, as they feed
on pondweed, duck potatoes,
and other aquatic plants.

mallard tames easily, and you can see it often on park lakes and streams. It feeds mostly on aquatic vegetation and grains—sometimes you will see them in farmers' fields—and it destroys large quantities of mosquito larvae. The black duck is one of the shyest, wildest of our dabblers; just as you think you have finally crept up on a pair for a better look, off they go swiftly, their sharp ears having caught the slightest sound you made. The pintail, a particularly handsome duck, is another you will often find in marshes all over the country. It is one of the most widely distributed North American ducks and is a hardy creature, being one of the first ducks to come north in the spring. It often arrives so early in northern marshes that there are still patches of ice on the marsh.

The other dabbling ducks behave much like the wood duck. They do not nest in trees, though, but prefer to hide their nests among the aquatic grasses, sedges, and shrubs. Their diet ranges through many kinds of submerged and emergent plants, a wide variety of insects and aquatic animals, as well as acorns, other nuts, and certain farm crops.

The diving ducks

Another group of ducks, those that dive and swim underwater, have shorter legs, placed farther to the rear of the

The ring-necked duck, often called the ringbill, travels in small flocks, alighting in the open water of ponds and streams near swamps but frequenting the shallows. They are good divers and can obtain their essentially vegetarian fare in deep water, if necessary. The ringbill's voice has an unusual purring sound.

body, than the dabblers'. These adaptations, though they make the divers awkward on land, give them fast propulsion and quick maneuverability underwater. Quick as a flash they dive at the slightest disturbance, surfacing scores of feet away in a direction you do not expect. Two common divers are the canvasback and redhead ducks. Though you will find them along our eastern coasts in winter, they nest in the pothole country. Both feed primarily on aquatic vegetation and seeds, though shellfish, insects, and fish form part of their diet. Both are considered good game and, unfortunately, loss of habitat and overhunting have reduced their numbers.

Another group of diving ducks you will recognize is the mergansers, the most highly adapted to water. Their streamlined bodies permit them to swim underwater at great speeds. Their bills are slender, unlike the flattened bills of most ducks, and they have toothed edges that help hold slippery fish. The American merganser is commonest on northern rivers, remaining wherever open water is available. It can eat great quantities of fish, devastating schools of whatever fishes are most abundant, including at times young salmon and other game fish. The red-breasted merganser of salt marshes also eats quantities of fish, but it goes after other small animals, such as crayfish, and small marine crustaceans. In summer the hooded merganser is the one most likely to be found on ponds and in marshes, since its diet is

Redheads are late migrants and fly in irregular flocks but join other diving ducks on water. They dive for food as much as twelve feet below the surface, primarily in search of aquatic plants. The male has a deep, vibrant call that has been compared to a violin tone or a cat's mewing.

small fish, frogs, tadpoles, and aquatic insects. Because these mergansers, like the other diving ducks, have legs placed well back on their bodies, it is more difficult for them than for the dabbling ducks to take flight from the water. They need to run across the surface, or skitter along, for a short distance before becoming airborne. Thus mergansers and other diving ducks prefer more open water than do the dabbling ducks. This is a satisfactory arrangement, since the larger bodies of water often support more fish than the sluggish warm waters of a small pond or marsh. The divers hold the small fishes in check, and as a result sport fishing is better because larger fishes have more food.

Geese, grebes, and coots

Geese belong to the same family as ducks and swans and, like all of this family, they breed and nest in the wetlands. You can recognize them because they are larger than most ducks, with less flattened bodies and longer necks. Predominantly vegetarian, they also eat mollusks and insects if plant food is insufficient. Their powerful legs, placed well forward, and their sturdy, webbed feet give them speed and agility in water and on dry land. They regularly forage on land for grasses and farm grain to supplement their diet of roots, tubers, wild celery, eelgrass, and other aquatic plants.

Snow geese are so gregarious that several flocks with a total of over forty thousand individuals have been seen together. The female usually incubates her eggs in a nest lined with down, and the male assists in the care and feeding of the young.

All our native geese nest principally in the northern states, Canada, and into the Arctic. Both in flight and on the water, they are among our most spectacular water birds. Geese are noisy when they fly; the sound of these honkers as they pass overhead in their striking V formation will never be forgotten by anyone who hears it. Huge flocks of snow geese and blue geese settle in the Gulf coastal marshes, and in their huge numbers they can do serious damage to the vegetation since they prefer the roots and underground parts of growing things. When residents of Texas and Louisiana shoot the birds, they are unknowingly helping to restore some of the balance.

Other water birds you will find in the marshes are the grebes and coots. Grebes resemble ducks in their feeding adaptations, though they do not belong to the duck family. Their legs, likewise short and set far to the rear, and their

The eared grebe's nest is a feat in engineering though untidy in appearance. The shallow nest forms a floating mass attached to nearby vegetation.

Six of the eighteen known species of grebes live in North America. During the mating season they have attractive head ornaments and perform elaborate dances in the water.

137

The toes of the American coot have broad pads that equip it to swim as easily and smoothly as a duck does. Like the diving ducks, it runs across the water for some distance before becoming airborne. Coots have a large "vocabulary" of sounds described as grunts, coughs, quacks, whistles, wails, croaks, toots, squawks, and chuckles.

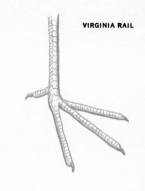

VIRGINIA RAIL

lobed feet give them maximum power in the water. Grebes are superb divers and swimmers, and their young begin to swim almost as soon as hatched. But grebes are clumsy on land and are unable to take to the air except by running along the surface of the water.

The coot, one of the friskiest and noisiest of our water birds, is related to the clapper rail. It feeds primarily on pondweeds and bulrushes on the surface. Coots swim with their heads bobbing and make a great splash and splatter when they land on water or leave it.

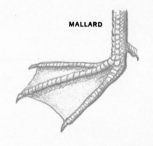

MALLARD

The elegant wading birds

Birds that stride into the water in pursuit of their prey or stand poised in its shallows until something edible swims near are known as the *waders*. They have spearlike bills that can be aimed accurately, and all have fairly long legs that allow them to hunt some distance offshore. American waders generally belong to the heron family.

The great blue heron, with its pale gray body, long neck, and slender dark legs, is stately and beautiful. Like most herons it nests and roosts on a platform of sticks, generally in colonies. The great blue heron feeds on several kinds of fishes, but it also destroys many predators of fish, such as water snakes and carnivorous fishes. It will stand motionless for long periods in shallow water as it waits for prey.

The snowy egret, another heron, is one of the most striking of all wading birds. It breeds as far north as New Jersey to Nebraska. In the United States, it is now protected from hunting and is most often found in the marshes of South Carolina, Georgia, Florida, Mississippi, Louisiana, and Texas, though it is occasionally seen in the summer as far north as Long Island and the Great Lakes. Once seen it is never forgotten. Totally white except for its black bill, black legs, and yellow feet, in the breeding season it develops a cape of plumes that sweeps over its back. Its larger

PIED-BILLED GREBE

The long, widespreading toes of the Virginia rail keep it from sinking into the wetland soil as it forages for food or runs to escape a predator. Ducks have webbed feet which serve as paddles when they swim. Dabblers such as mallards also use their feet to upend themselves in the water while they pull food from underneath. As the pied-billed grebe swims, the stiff, horny flaps that line the sides of its toes fold back as the feet push forward and flatten out on the return stroke.

139

Herons and egrets fish for their meals in the shallow waters of marshes and swamps, employing a number of virtuoso techniques to catch their elusive prey. The great white heron (1) uses the *stand and wait* method: it stands motionless until a fish swims by, then snaps it up. The little green heron (6) uses essentially the same technique, but in more shallow water. *Canopy feeding* is employed by the reddish egret (2): it dashes about until a frightened fish seeks refuge in the shadow cast by its outspread wings. All herons and egrets use the *walk and wade* method from time to time, but the most adept practitioner of this technique is the common egret (3): the bird simply walks . . .

cousin, the American egret, is also snow-white and has black legs and feet, but it has a yellow bill. This magnificent heron is seen often in wooded swamps and in fresh-water as well as salt-water marshes. In the early 1900s both of these birds were almost exterminated by plume hunters.

Green herons nest on platforms made of sticks, which they usually build in low trees or shrubs some distance from the water. Common and widespread, they nest alone or in colonies. They are greenish blue in color and have bright orange legs that are shorter than those of most herons. If you hear a peculiar cry, a *skeow*, that sounds like the noises you make by blowing on a blade of grass held tightly between your thumbs, you can be certain a green heron is nearby.

The two most inconspicuous waders on the marsh are the bitterns, whose relatively short legs restrict their hunting range. Until you know it well, you could search for hours

without seeing a single American bittern. It is known as the "stake driver" because its most typical call resembles the sound of a stake being driven into the ground. This species and its diminutive relative, the least bittern, feed on a wide variety of small aquatic creatures. Bitterns breed over an extensive region, from Canada to the southernmost marshes of California, Arizona, Kansas, and New Jersey. For many centuries, the Indians used bittern bills as arrowheads.

All these elegant waders can be seen in both fresh- and salt-water marshes over a vast area of North America. Ducks, geese, coots, grebes, rails, shore birds, and these beautiful waders are part of the abundant bird life typical of the wetlands. They give marshes and swamps their character. Here they find a rich and varied diet of aquatic plants and animals. They are dependent upon these wetlands; without them they would disappear.

. . . the shallows, searching for food. Probably the most acrobatic method of feeding is *pirouetting*, used by the Louisiana heron (4). This bird steps forward, wings extended, and turns in place. It raises one wing and sticks its head underneath to look into the water. As the heron completes its turn, it raises the other wing and takes another look. The slow, erratic shadows cast by the wings help to lure the prey into the trap. The snowy egret (5) hovers in midflight, then stirs the water with its foot to flush out its meal. Whatever the technique, each heron and egret stakes out its private feeding territory and defends it against all comers.

This snowy egret has grown about fifty threaded, lacelike white plumes, or aigrettes, on its back between the wings. They reach the height of their beauty in the breeding season. During the late nineteenth century the beautiful plumes were so fashionable that thousands of the breeding birds were killed every year by plume hunters, who thus threatened the egrets with extinction. Because egrets nest in colonies, or "bird cities," the hunters had only to track the birds by watching the direction of flight to and from the feeding grounds. After shooting the parent birds and removing the valued plumes, the hunters left the young to starve.

Because of the growing alarm at the increasing rate at which plumed and other valuable wild birds were being killed, the Audubon Movement began. After years of work, the National Audubon Society succeeded in having laws passed to protect the birds, and sanctuaries were established. Still, the plume hunting went on. Before it could be stopped, two Audubon wardens were killed and others wounded by hunters they were attempting to arrest.

In the years since the snowy egret has been protected, the bird population has increased to tens of thousands and the species has been saved from extinction.

Fur coats for mammals

You have seen the variety and abundance of plant and animal life in the wetlands during the productive months of summer, as well as some of the ways plants and animals are adapted to this environment of water and land so that they can utilize the protection afforded and the food offered by these two worlds. But what happens when the big freeze sets in? What happens when the plants die back and ice chokes the waterways? The open marsh swept by the bitter winds of winter is a particularly harsh environment. How do the animals that live through this season survive?

In mammals the development of hair is one of the commonest adaptations to cold weather. Hair consists of non-living cells that developed under the skin and were pushed out from hair follicles. A dense growth of hair—a *fur*—provides insulation, reducing the loss of heated air next to the skin and retarding the penetration of the cold air from without. Fur keeps an animal's immediate environment close to body temperature.

The hair in fur is of two varieties: the soft, thick underhair that lies next to the skin and the coarse, shaggy guard hair that projects through it. This outer coat, which in different mammals varies in density and depth, protects the underhair from wear and tear. If the underhair is damaged, the fur becomes much less effective as a means of insulation; and, if the damage is extensive, the animal may die.

Of the numerous fur-bearing mammals that shelter in the North American wetlands, three are especially notable. Muskrats, beavers, and mink live from the edge of the arctic tree line through Canada and most parts of the United States down to the Gulf of Mexico. All three are excellent swimmers and show, in varying degrees, adaptations for living in and near water.

The fur of mammals is an insulating cover that helps to retain body heat. It is kept waterproof by oil secreted from the animal's glands. The underhair, a dense, thick, and soft layer, lies next to the skin. The guard hair, which is longer and coarser, forms a protective outer layer. When the pelts of fur-bearing animals are processed for furriers, these guard hairs are plucked out.

Beavers—wetland woodcutters

You have to be lucky to see beavers, since most of their work is done after dark. And even if they are about during the day, your approach will alert the entire beaver colony and bring about tail-slapping danger signals, and these large rodents will disappear underwater.

144

In ten minutes, a beaver's two large chisellike front teeth, or incisors, can chip away enough wood to fell a sapling six inches in diameter. When the beaver senses that the tree is about to fall, it scrambles to safety, returning only seconds later to begin cutting up the branches.

BEAVERS: WETLAND ENGINEERS

The night-working beaver carries construction materials through the water to a damsite. . . .

. . . Emerging from the water, the beaver heads for a break in the wall of the dam. . . .

*. . . Arriving at the break, the beaver
will use its forefeet and teeth to
wedge sticks in place.*

Beavers are the construction crews of the wetlands, undoing old
landscapes and creating new ones. Once they have chosen a damsite,
they are hard to discourage. Lighted lanterns, burning sulfur, even
moth balls have been placed on the dynamited ruins of their dams,
but the beavers push the gimmicks overboard and begin again.
The dam may back up water into nearby forests, cultivated fields,
and roadways, often to such an extent that the animals have to be
livetrapped (*right*) and moved to an area where they can carry on
their work to the benefit of beaver as well as man. In most cases,
however, their work is of great value. The marshy habitat they
create becomes in time a home for many other animals. Fishes
move in, as well as herons, muskrats, and minks. The beavers move
on after they have cleared the area of the trees they feed on, but
behind them a lush meadow develops, and with it more plant and
animal life appears.

Beavers may grow to about two or three feet in length and weigh from thirty to sixty pounds. Their hind feet are big and well webbed. The beaver's tail, long as in most of the rodent family, is naked, broad, and conspicuously flattened. Sometimes a beaver sculls with its tail, but it usually swims with alternate strokes of its powerful hind feet. Exclusively herbivorous, it feeds on the stems, roots, and foliage of plants in summer and on the bark of trees during the other seasons. Poplar or aspen bark is its preference, but other trees will do as well. It eats three pounds of bark in an average meal. The beaver's *incisors*, two pairs of protruding front teeth, act like chisels in stripping and cutting. They are constantly worn down but never stop growing, and they last a lifetime.

Beavers mate in midwinter and produce each spring a litter of three or four baby beavers called kits. The female nurses her young for six weeks, though they begin to eat plants, leaves, and bark morsels by the end of their first month. By midsummer the kits reach a weight of eight pounds. Parents, kits, and the offspring of the two preceding years form a close cooperative social unit called a *colony*. The parents control the size of their colony rigorously, never permitting it to exceed about fourteen members. Each year the mature two-year-olds leave or are expelled and found colonies of their own, sometimes traveling thirty miles before settling down.

Building a home

The industry of beavers has become proverbial. In all countries where they are known, people speak of "busy

beavers." Diligently at work from midsummer till the approach of cold weather, they perform feats of engineering and construction unparalleled by other mammals except human beings. Their labors—arduous, diversified, and interrupted only for brief rests—have a single end: to provide shelter and food for the winter months ahead. The climax of this work, their masterpiece, is the *beaver lodge*.

As winter nears, all members of the beaver colony carry in their forepaws loads of mud and sod, decayed leaves, and other material to plaster the interlocked twigs of the lodge walls. When the cold comes, these lodges freeze solid, conserving the warmth of the water below them and the body heat of the occupants. The stout icebound lodge walls often can withstand even the mighty blows of a marauding bear. Nevertheless, even when they are drifted with snow, such igloos obtain sufficient ventilation. The warm air inside rises through the dome and melts the snow above it into a tiny inverted cone, through which the cold, heavy fresh air flows down in a small continuous stream.

Beavers change the landscape

Beavers can perform much more elaborate feats of "engineering." When their environment is unsuitable for colonization, they often transform it. Swamps, with their muddy surfaces and occasional shallow streams, seldom contain water of sufficient depths to accommodate beavers. When there is extended cold, the streams freeze to the bottom. A family that built its lodge in such shallow water would risk being cut off from its food supply.

Usually the colonizing beavers choose wetlands through

Beavers spend the winter in lodges constructed during the warmer months. These dwellings average seven or eight feet in diameter, large enough usually to accommodate a colony of fourteen. Two or more plunge holes in a lodge assure access to the water under the ice and to the food supply stored there during the summer and fall. The food is the nutrient-rich inner bark of trees. The interior of the lodge has a main floor some four inches above the waterline and sleeping shelves made of soft shredded wood. When beavers build this domelike structure, they usually choose a site that will get the most sunlight during the cold months. They begin by raising a mound of sticks and piling on mud to serve as cement. A venthole is left on top. Inside the mound the builders tunnel out their living quarters. When the world outside is frozen and barren, the beaver has a warm, ventilated home and an abundant food supply within swimming distance.

Whatever the reasons beavers build dams, these structures do raise the water level, thus making it possible for the animals to forage underwater for food in winter. The building materials of the dam include sections of trees, aquatic plants, stones, and mud.

which streams flow, and build dams to back up the water on the stream and increase the depth. These beaver dams are often low and sometimes only a few yards long, but they have been known to reach a height of twelve feet and a length of half a mile or even more! When building dams of some magnitude, the beavers lay heavy cuttings of trees on the stream bed, meticulously placing the large, relatively immovable butts so they face the current. Stones weighing up to six pounds may be piled on top of the timber, securing it still further. Branches and brush are piled up and stopped with mud, leaves, and other debris. Year after year the beavers may extend these dams wider and higher. The impounded water may spread over hundreds of acres, supporting ever-increasing quantities of plants and animals.

The muskrat

The muskrat, a relative of the beaver, is equally aquatic but much smaller, rarely reaching three pounds. With its plump little body, whiskered face, and long naked eight-inch tail, it looks like an overgrown mouse. Its tail, unlike the paddlelike one of the beaver, is flattened on the sides and is quite scaly. The hind feet of this rodent are partially webbed, thus giving them additional thrust when swimming.

150

The soft upper fur of the muskrat is a uniform dark brown, in contrast to its gray underparts.

As is generally the case with small mammals, the muskrat mates repeatedly and promiscuously. The female may produce three litters a year, beginning in early spring. Males that claim the same mate inflict severe, sometimes fatal, wounds on each other.

Two little scent glands, present in beavers as well, reach their fullest development in muskrats and give the species its name. The glands, or *castor bags*, produce a reeking musklike odor, especially strong at breeding time. Males and females alike make a nightly circuit of feeding platforms, beaver dams, islands, and mussel banks, spraying these resorts of the muskrat clan with their powerful perfume. This ritual may help to establish easily located breeding sites and makes reproduction less dependent on accidental encounters.

Muskrat lodges vary in size. Some of the biggest have been known to be ten feet in diameter and four feet high, and the smallest hardly larger than a bushel basket. They can be based on a dense mass of sweet flag, a group of willow sprouts, or a pool choked up with emergent plants and silt. Most often, the muskrats choose a moist little promontory and have to dig a channel and plunge hole several feet below the floor to reach navigable water.

The muskrat feeds mainly on plant material. In the cattail marshes of the East, it consumes large amounts of arrowhead, pondweed, bulrush, bur reed, and clover as well as water lilies. In the coastal marshes of Louisiana, where the muskrat populations are sometimes huge, its chief plant foods are three-square bulrush and paille-fine grass. In California it favors the sizable piles of seeds borne by water and wind to the shore. But wherever muskrats live, they dote on cattails. In some areas, these rodents often do such severe damage to cattail marshes that heavy trapping must be carried on in order to keep the muskrat population within manageable numbers.

Muskrats sometimes add meat to their diet. They have been seen eating fish, clams, insects, and snails. The muskrat especially enjoys mussels and regularly explores the banks and bottoms of lakes to procure them. When winter is retreating and cracks form in lake ice, one of the little rodents now and then moves out on the lake surface, dives through a crack, and soon emerges from the water with a mussel in its mouth.

Though awkward on land, muskrats are expert and agile swimmers that use their vertically flattened tails as rudders. In warm weather they forage for marsh plants and for mussels and other invertebrates. As fall approaches, they set about building feeding platforms from which they set out in winter to search for food.

The base the muskrat builds for winter is a lodge like this one, or a burrow in a marsh bank. Like the beaver, the muskrat prefers to hole up in a weatherproof retreat during winter. Moving in and out of the lodge's plunge hole, the animal patrols the underwater pasture for its food.

The mink

The mink is a slim, elongated member of the weasel tribe. Male minks are seldom much more than two feet long and weigh about two pounds. The female, considerably smaller, rarely reaches one and a half pounds in weight. Both have dark, rich brown fur, with a patch of white under the chin. Except for their larger size, minks closely resemble weasels, but they are much more fond of water. Even so, minks are not so closely bound to the water as are otters, also members of the weasel family, which often travel some distance overland between watercourses.

Minks are exclusively carnivorous. They build their dens in wooded areas near water. Primarily nocturnal, they patrol nearby marshes and waterways, looking for food. They have powerful, well-developed canine teeth, and when they seize a victim it seldom escapes. Throughout the year they kill and eat insects, frogs, fishes, muskrats, and other aquatic life.

Minks, along with other carnivores, help keep these smaller animals at a favorable level in and around the marsh.

The mink is an agile swimmer and diver. It has few enemies, although a fox, bobcat, or large owl may occasionally kill one. The trapper has been its main predator.

Feeding in the autumn marsh

Fall is a time of constant stir and activity in the wetlands. The vegetation is crowded with birds in transit, alighting to drink and feed, and with other birds that have arrived for a longer stay. The marsh hawks that circle overhead live principally on rodents, but they and the sharp-shinned hawks are diversified hunters. These hawks are looking for young rabbits, rodents, birds, and other prey in the marsh below. Mink of all ages prowl the crayfish-rich zones close to shore and sometimes hide in abandoned muskrat lodges from which they can lunge out to waylay a passing grebe or coot. Occasionally they assault young beavers that have been rejected by their parents and have not yet learned the wariness essential on the marsh. Muskrats, too, are often victimized. But so long as food of all sorts is available to minks, they do not prey heavily on their furry neighbors. When they do, the inexperienced, the outcast, and the disabled are their victims. The minks divide their attention between water, shore, and wooded uplands and may range many miles in a day.

The approach of winter

In the northern wetlands the days are now darkening early. There is a chill bite in the morning air. Even in the middle of Indian summer ice may form on the ponds each night, and where the water is shaded and quiet the evening ice may remain until the next afternoon. The change of seasons brings about a change in the behavior of the inhabitants of the marsh as they prepare for the winter days ahead.

Under cover of night the beaver is occupied with felling trees, repairing its dam, and building its lodge. At dusk, beavers all over the marsh set about the construction that will protect them during the harsher times ahead. They work strenuously, rarely pausing, ears strained to detect the approach of an intruder. The snap of a twig nearby or the

A mink, having killed a muskrat, may devour it on the spot or drag the corpse to a den where several other of its prey may be stored. This member of the weasel family feeds on a variety of marsh animals, including even swiftly swimming fish.

153

rustle of sedge on a windless night stirs them to instantaneous action. One after another they crack their flat tails down sharply on the surface of the water, then dive to safety. The sounds they make snap out like rifle shots across the dark marsh. Their action when alarmed is probably a reflex, but it serves to protect the whole population.

At last there comes a night when the mud develops a permanent crust and the rocks on the shore turn white with frost. The water surface becomes frozen except where birds were sitting. By morning many thousands of them are gone. The grebes and coots and mild-weather ducks are miles away, heading south where the ice will not arrive till later or not at all. Mallards and green-winged teals, however, remain and are joined by many newcomers from farther north. Mallards are rugged and do not mind the ice provided there are cornfields nearby. They can keep an area in the water free of ice by their activity. Loons also stay on, sometimes well into the winter. If there are open spaces in the ice, they can dive into one, capture a fish, and come out through the next patch of open water.

Among the late-staying birds are bitterns, pintails, and

the mergansers that may stay as long as there is a remnant of open water.

When the tight freeze comes, the last remaining birds depart. They wheel upward by the thousand, and the infection spreads. Wave after wave passes overhead a quarter of a mile from the ground in characteristic formations. Their departure calls are followed by months of long silence. Only now and then in the bitter cold does a circling crow or raven emit its harsh, melancholy cry.

The grasses and sedges have withered; the rigid cattails stand bare. All over the marsh the newly built lodges of beavers and muskrats become more readily visible, but the architects are not to be seen. They have settled inside until spring returns.

The bitter cold of winter

Few animals are safe in the bitter cold that has enveloped the marsh. The swift-moving mink may now feed on dead animals, or *carrion*. Its nose can detect the odor of death.

A flock of mallards takes refuge amidst the reedy stubble of a marsh in winter. These ducks often go no farther south in winter than is necessary to find open water. They return northward as soon as the ice melts.

A hungry snowshoe hare stretches as far as possible above the snowy surface to nibble bark. Winter must be nearly over, for this hare's white fur is beginning the change to its summer brown shade. This alteration in fur color is triggered by seasonal variations in daylength.

Its sharp toes are good enough at digging to break through the frozen shrouds of pheasants, quail, moles, shrews, and rabbits. Passing a muskrat lodge, the mink senses the presence of the healthy occupants. At times it is encouraged to batter and scrabble at one of these shelters. It attacks the southern wall, which faces the sun and is thinnest because ice and snow have not accumulated there. If the muskrats have not built firmly, the mink can finally get a paw in. Soon the hole is widened enough to admit the slim mink's weasellike body. More often, however, it must rely on the carcasses of animals it has killed and cached in holes.

Comfortable under the ice, contented muskrats may in time become available to the minks. It is mainly a question of weather.

Freeze-out

The ice that protects aquatic creatures can in time become a deathtrap. Although it excludes predators and those animals that eat the same foods as water-dwellers, thick and snow-covered ice also keeps light and oxygen from entering the water. When photosynthesis stops, there is a fixed supply of oxygen dissolved in this water under the ice. Unless the ice cracks and opens up, no more oxygen can get in.

Until spring returns, fishes, salamanders, mollusks, crayfish, water boatmen, and snails must live on this slowly diminishing oxygen supply. So must the plants. Week by week they take in oxygen and use it in the cellular respiration that keeps them alive. Week by week they pour back into the water a waste gas, carbon dioxide. Only the submergent green plants add excess oxygen beyond their needs. All the animal organisms that keep alive under the ice use oxygen and deplete the limited supply of this gas. They also release wastes that in concentration will make the water uninhabitable. The decomposers further poison the living space with methane and hydrogen sulfide. By midwinter it is hard for the fishes to get enough oxygen.

A coating of ice glazes the withered stalks of cattails in a winter marsh. The thick, brown, fuzzy spike that tops the cattail contains hundreds of minute, one-seeded fruits that may find a suitable place in which to germinate when spring arrives.

156

fishes frozen in a plunge hole. They fight weakly with one another and gnaw the remains of other muskrats now littering the ice. Tails and feet frozen stiff, eyes misting, wandering in circles, they barely see the mink or the great horned owl as it glides across the icy barrens toward them.

All the marsh animals need to obtain the special foods suited to them. In a particular winter minks may eat a good number of muskrats to stay alive. The muskrats eat mussels as well as .plants. The beaver needs bark. Whatever their diet, they must depend on a liberal food supply to thrive and maintain healthy populations.

A good many muskrats may die to keep the mink alive. That is the way of the marsh. From another point of view, the mink is beneficial. It helps keep muskrats from dominating the marsh, killing off all its plants and eventually driving out the other animals that need plants.

The warmth of spring

The warming rays of the sun at last return to the marsh. The intense activity and productivity of spring and summer begin again. Plants grow rapidly, and animals, taking advantage of more plentiful food, find mates and replenish their numbers after the rigors of winter. Plants and animals, of course, are adapted to the conditions of any niche in any habitat, but you have seen how there is a particularly rich variety of life in the fresh- and salt-water marshes where land and water continually interplay. The story is much the same for swamps and bogs. All these kinds of wetlands have a rich and varied animal and plant life. It would seem that without any difficult stretch of the imagination man could see how valuable the wetlands are to himself as well as to the larger natural world around him.

Throughout millions of years every plant and animal species that survives in a wetland has developed its own unique adaptations. The bald cypress, which thrives in swamps, is one of the trees that evolved in ways suitable to the conditions of their habitat. Others are black willow, red maple, green ash, and sweet bay.

Wetlands
or Wastelands?

Marshes and swamps have too often been considered wastelands, valueless expanses of water and sticky mud unfit for farming or any other use. Since Colonial days man has set himself to the task of reclaiming them. Wetlands have been ditched and drained, or filled in with thousands of tons of earth or tin cans, old cars, and refrigerators, and almost overnight a new housing development, industrial plant, or farmland has been created on them. Some small countries with limited land, such as Holland and Denmark, have had to reclaim their salt marshes for land on which to produce food. But we have vast lands in our country, and we have had no food shortages; yet we have completely destroyed many millions of acres of wetlands by filling. In most cases filling has caused an unnecessary and irreversible loss.

In prairie pothole country, marshes have been drained to make way for cropland. This land reclamation has in many instances been performed hastily and unwisely. Many potholes that were drained could not be put to economic use, and the conversion often had a disastrous effect on the water supply and soil quality of the surrounding agricultural region. Drainage, wherever it was practiced, greatly raised the fire hazard. And in the flyway areas, where it was practiced most commonly, the destruction of bird and animal breeding and feeding grounds drastically reduced an important national asset. Ironically, many of the drained marshes have been allowed to revert to their natural state.

Are such failures in land reclamation sufficient reason to save or "re-create" these vast tracts of soggy real estate? Salt marshes do, after all, serve as breeding places for mosquitoes, and they are filled with black sticky silt that might discourage most people at the outset from visiting them. To many people a coastal marsh is not nearly so attractive as a forest. Few people would fail to be aroused if a proposal were made to cut down the remaining coast redwood forests in California or to sell Yellowstone National Park to land

This garbage dump has displaced the fish and waterfowl that lived in the productive marsh once located here. The destruction of this wetland reflects what is happening throughout the United States: 127 million acres of marshes, bogs, and swamps have been reduced to about 74 million acres, and these are disappearing at a rate of one percent each year.

speculators. But few object when a city destroys its marshes by covering them with garbage or junk.

The idea of productivity

One reason for protecting our wetlands lies in their biological *productivity*. Productivity in marshes, as in any ecosystem, is its "output" in terms of living things—plants, marsh crabs, ducks, muskrats, to name a few—brought about by the interaction of the community in this particular environment.

Productivity is measured on the basis of three factors. First, you must determine the total quantity of living material present in the marsh at a given time. This is called the

standing crop, or biomass, generally expressed as the dry weight of all living matter produced per given area of habitat. Second, you must know how fast the standing crop is being removed, that is, how many wild rice plants have been eaten by ducks, how many ducks have flown south or have been killed by hunters, and so on. Third, you must measure the rate at which these same members of the natural community are being replaced—the quantity of wild rice plants growing each year, the number of ducklings born and the number surviving their first year.

In contrast to some other valuable resources, such as oil wells or mines, a properly protected marsh cannot be depleted. It continually produces, like the legendary pitcher of wine that is never emptied. And, in a sense, marshes are economical producers: unlike a farm they do not require vast sums of money to make them "work." But are they valuable? Do they produce efficiently, as a successful farm does?

The answer is yes. Tidal marshes are among the most productive lands in the world. They are four times more productive—that is, they produce four times more plant growth—than a good cornfield. They are surpassed only by the sugar-cane fields of Hawaii. Another point to remember is that both corn and sugar cane require heavy application of fertilizers, but no one fertilizes a tidal marsh! The crop from a field of corn or sugar cane is readily visible since it is measured directly in numbers of bushels or tons harvested per acre. In contrast, the tidal marsh does not yield its crop directly to man. You can see it only as a reflection in the abundance of finfish and shellfish in our coastal waters, in the number of ducks in the pothole marshes of the Dakotas or the vast marshes of the Klamath Basin, and so on.

The nutrient trap

Recently ecologists have turned their attention to the scientific study of productivity in marshes. As a result, some interesting facts have been discovered. Measurements in Delaware Bay show that there is an unusually high concentration of nitrogen and phosphorus compounds as well as carbohydrates and other nutrients in estuarine water. Why is this so? A certain portion of the standing crop of marsh

Deep oceans and deserts or semi-arid land cover a very large proportion of the earth's surface, and yet these environments have a very low productivity level. In contrast, the wetlands, comprising only a small percentage of the earth, are among the most productive of all natural habitats.

OCEANS

WETLANDS

DESERTS

GRASSLANDS, FORESTS, AND LAKES

0   50   100   150   200

POUNDS PER ACRE PER DAY

166

This aerial photograph of Cape Henlopen shows the contrast between the calm estuarine waters of Delaware Bay on the right and the more turbulent ocean waters on the left. Estuaries and bays like this are the breeding and nursery grounds for many of the fishes harvested commercially.

plants dies each year. Organic and inorganic nutrients are once again fed out into the bay water. In addition, the mud-banks are covered with a brownish-yellow film of diatoms, and on the marsh surface amidst the grasses one often finds snatches of filamentous algae. These inconspicuous, often overlooked forms are tremendous stores of energy, yielding sugars and oils.

The abundant nutrients are kept from being lost to the sea by a "nutrient trap." You can probably guess what happens in an estuary where the heavier salt water meets the lighter fresh water. The fresh water, flowing down from the upland, tends to float on top of the salt water. In the estuary where this occurs, the nutrient-filled water is held back from the sea and is then recirculated. Twice a day rising ocean tides send the trapped nutrients surging back over the marsh. The result is a constant bath of nutrients: flood tides surge landward, bringing nutrients to the marsh, and ebb tides flow seaward, returning nutrients to the bay.

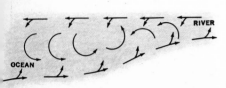

In an estuary, the lighter fresh river water flows over the heavier ocean water and tends to remain on top of it. The salt water moving in underneath creates a somewhat circular motion that forms the "nutrient trap."

How stored nutrients are used

Plants use nutrients, such as nitrates and phosphates, mostly in an inorganic form. Bacteria are chiefly responsible for the breakdown of organic nutrients into inorganic ones. Since Delaware Bay is shallow, the water is constantly circulated and mixed. Because of the large quantities of silt the waters often get murky, and so the penetration of light is limited. Some of the inorganic nutrients are transported upward by the mixing water into the zone of light penetration—the *photic zone*—where they are utilized by algae, especially diatoms and dinoflagellates, in the process of making new protoplasm. Other nutrients are also used beneath this zone by bacteria and other organisms that do not need sunlight to live.

Animals, however, require most of their nutrients in an organic form. You know that sugar, for example, is a basic food, and a tidal marsh can produce an astonishing amount. A marine biologist, Carl N. Schuster, Jr., estimates that over 500 pounds of sugars may be produced on a single acre in the Delaware tidal marshes each year. He has calculated that if only half of Delaware's 130,000 acres of marshes produced sugars at this level, there would be an annual crop of over 32,500,000 pounds for the fishes, clams, oysters, and other marine life in the surrounding waters. At 10 cents a pound for sugar, the tidal marshes would be worth over $3,250,000. It is also estimated that as the bacteria break down the sugar-producing plants, 18 pounds of phosphate minerals are released each year in every 50 gallons of water flushing over the marshes. This means that in every 50 gallons of water there would accumulate 18 pounds of phosphate fertilizer available for use by floating microscopic plankton, mud algae, and marsh grasses; these, in turn, make more sugar that will feed more clams and oysters in the endless cycle.

Along the Atlantic Coast, cordgrasses are produced in enormous quantities. On one acre of Georgia salt marsh 4.8 tons of these common marsh plants are produced each year. High production is also found in Canary Creek Marsh in Delaware. Each year this marsh produces 323 tons (dry weight) of plants, mostly cordgrasses, on its 123 acres. It is estimated that 84 tons of this material, in the form of detritus, are carried out of the marsh by the tide each year.

The constant bath of nutrients carried by tides surging in and out of tidal marshes *(dark arrow)* results in a very high biomass. This productivity is of direct importance to man, who is the recipient of the marshes' wealth, not only in the fishes and shellfish shown here, but in muskrats, waterfowl, and other products of this wetland habitat as well.

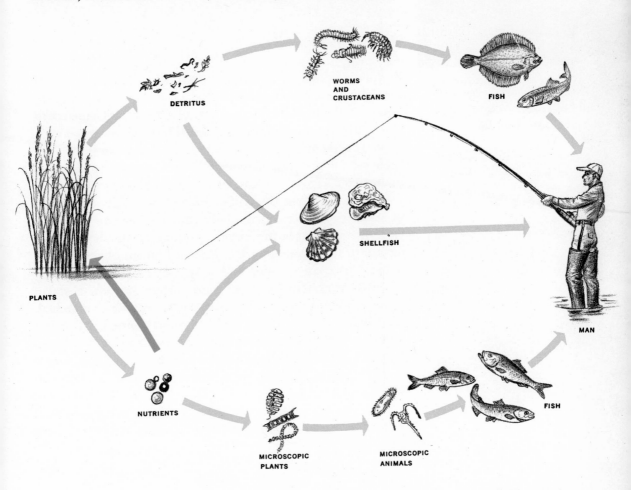

DETRITUS

WORMS AND CRUSTACEANS

FISH

SHELLFISH

PLANTS

MAN

NUTRIENTS

MICROSCOPIC PLANTS

MICROSCOPIC ANIMALS

FISH

The role of the ribbed mussel

In addition to the physical trapping of nutrients by the movement of water, important *living* traps exist in estuaries. One fascinating living nutrient trap is the ribbed mussel, embedded by millions in the marsh mud. They filter huge quantities of water, straining out some of the plankton and energy-bearing detritus trapped in the estuarine waters flowing over their gills. But large quantities of detritus accumulate on their gill surfaces, and the mussels expel the material periodically by opening their shells. These organically rich so-called pseudofeces build up as sediment and are available as nutrients for plants such as algae, which in turn support animals such as fiddler crabs.

Studies have shown that these mussels contribute to the high level of phosphorus in the waters of the estuary. Furthermore you can see that they are important in supplying food that ultimately supports fish and other shellfish in the bay, such as clams and oysters.

Shellfish and finned fish

Because of the high nutrient content of a tidal marsh, plants and animals are abundant, and some are of great value to man. At the mouth of the Niantic River in Connecticut about fifteen thousand bushels of scallops are harvested each year. This amounts to about three hundred pounds per acre each year, more protein than you can produce in the form of beef on a good pasture of the same size!

Oystermen along the Atlantic Coast regularly dump tons of clean oyster and clam shells into coastal bays to serve as attaching places for larval oysters. Oysters hatch from eggs into free-swimming larvae, which swim about for two weeks. After an eyespot and a special foot develop, they begin to search for a clean, hard surface—a rock, a stake, a shell—to which they can attach themselves. Once attached, the young oysters, called *spat*, become nonmoving, sessile animals. The spat prefer to find a hard surface in the down-bay area, where they are fattened by the bath of nutrients in the form of detritus brought by the ebb tide from the marsh.

Our coastal fisheries are greatly dependent upon tidal marshes. Menhaden, striped bass, bluefish, and flounder lay

Each two- to three-inch-long ribbed mussel pulls in and recirculates about four quarts of water every hour. In an estuary, where these mollusks live in incredible numbers, they are responsible for the movement of millions of gallons a day.

Oysters are harvested from shallow-water beds located
in areas with moderate wave action and where the
salt content of seawater has been reduced by the flow
from tributary streams. Oyster farming, which brings
in more than 28 million dollars a year, is the second
most valuable shellfish industry in the United States,
surpassed only by commercial shrimp fishing.

their eggs in the shallow waters of the estuary, which serve as nursery grounds. The menhaden feed primarily on microscopic algae and detritus; they use the delicate gill rakers at the backs of their mouths to strain out tiny diatoms and other floating food particles. Because these small fish are so oily, they are seldom sold as food, but the leather in your shoes may have been tanned with menhaden oil. It is also used in making paint, varnish, ink, insect spray, and even soap. After the oil is extracted, the rest of the fish is used in the manufacture of fertilizer as well as fish meal for chicken and hog feeds. These fish are thus an integral part of the food chain that produces the chicken and ham that you enjoy so much. Menhaden are found from Nova Scotia to Brazil, and the catch in the United States, larger than the catch of any other fish along our coast, totals 600,000,000 pounds each year.

Have you ever eaten fillet of sole? It is actually one of the flatfish, or flounders. Each year United States fishermen land 45,000,000 pounds of flounder and related flatfish. Along the New England coast some 12,000,000 pounds of winter flounders are caught each year in the shallow coastal waters. Unlike most fishes, winter flounders swim on their side along the bottom, though they begin their life swimming like regular fishes. Unless you have caught them, you are probably unaware that the adult flounders have *both* eyes on

Its flat shape, speckled black pattern, and grayish-brown coloring have camouflaged this flounder so well that it is almost invisible against the background of sand and pebbles. In addition, the flounder "shivers," thus throwing sand over its body until only eyes and mouth are exposed. Then, when a small shrimp, crab, or worm comes by, the fish darts up with remarkable swiftness to seize it. If a flounder misses the prey, its forward momentum may carry it two feet out of the water.

the upper side, the area of greater visibility. When they are five to seven weeks old, before the soft cartilage of the skull has hardened into bone, one eye begins to move upward around the head. The flounder's lower side, the one next to the bottom, becomes white, and its upper side a grayish brown—a perfect camouflage.

The food web supporting our coastal fisheries begins on the tidal marsh. The algae and detritus washed off with the changing tides nourish the shellfish. In addition, they feed tiny shrimplike crustaceans and the small fish, such as menhaden, that are consumed by flounder, striped bass, and bluefish; these, in turn, support our spectacular game fish—the tuna, marlin, sailfish, and swordfish that so many people enjoy catching. The estuarine marshes of southern Florida, especially those in Everglades National Park, support the main part of the Dry Tortugas' shrimp fisheries in the Gulf of Mexico.

Coastal marshes are important to the survival of some of our most spectacular game fishes—the marlin, for example. These huge game fish, weighing up to one thousand pounds each, may leap from the water as often as forty times in attempts to dislodge the fisherman's hook. Because of their soft jaws they are often successful.

173

A hundred years ago the semiannual migration of the whooping crane was a stunning spectacle. Hundreds of the great white birds filled the skies, and the countryside rang with their loud, clear buglelike cries. Now, even with great effort by conservationists, only forty-four survive. These almost extinct birds are a casualty of the disappearing marshes of our Great Plains. Only two possible habitats on the entire continent of North America are large enough for them: a summer nesting area at Wood Buffalo Park, in Northwest Territories, Canada, and a wintering ground in the Aransas National Wildlife Refuge in Texas. Every year each pair of these four-foot-long birds still stakes out its private territory of at least four hundred acres, which it defends bitterly against any invading cranes.

In order to preserve the few remaining whooping cranes, North America's largest migratory birds, hunters along their 2500-mile migration route, taken each October and April, are asked not to shoot any large white birds flying overhead.

Reduced coastal marshes mean reduced business

Our commercial fisheries are big business. In 1960 the New England fleet alone landed over 800,000,000 pounds of fish worth $61,500,000. About 500,000,000 pounds included fish that, like the flounder, live close to the shore and are fed by the food chain that begins in the marshes.

One of the main reasons for the recent decline in oyster production on Long Island is undoubtedly the increased filling and dredging of marshes in the area. The oysters no longer have the steady bath of nutrients from the marshlands. All of us realize that as our cities continue to grow, more and more people will need homes—but our tidal marshes are surely not the place for them. Space can be provided elsewhere by careful and intelligent planning. Unlike our fresh-water marshes, tidal marshes are lost forever once they are destroyed. Even in fresh-water marshes wildlife is threatened by drainage for agricultural and housing uses.

Duck factories in the pothole country

Our pothole country is the most important waterfowl nesting area in the United States. Less than a hundred years ago, probably fifteen million ducks were hatched here annually. Because drainage has reduced the number of potholes where

The North American prairie pothole region covers some 300,000 square miles, primarily in south-central Canada, northern Montana, the Dakotas, and western Minnesota. The area was formed about eleven thousand years ago, when glaciers of the Wisconsin Ice Age receded, molding the landscape into a poorly drained terrain pock-marked with millions of relatively small depressions, the potholes.

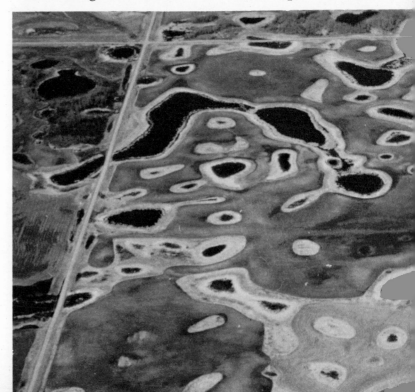

the birds once nested, now only about five million ducks are produced in the same region. A record year for drainage occurred in 1949–1950. Some 64,000 potholes, or 188,000 acres of good nesting territory for ducks, were drained in Minnesota and North and South Dakota.

There are potholes of all shapes and sizes. Some may cover more than a hundred acres, but most are small, covering less than ten acres. It was to these miniature marshes that millions of ducks used to return each spring from their wintering grounds farther south. Sometimes, old-timers say, the sky became dark as the ducks passed overhead on their journey along the central and Mississippi flyways.

Some pothole marshes are shallow depressions that hold water for only a few weeks in the spring. Others are completely covered with emergent vegetation and may hold as much as forty-eight inches of water in all but the driest years. Still other marshes contain large patches of open water ringed with marsh plants.

Ducks need various kinds of marshes at different stages during the breeding season. A duck selects a nesting site that is isolated from those of other ducks of the same species. Often the nest is built on dry ground, occasionally some distance from water. During courtship and breeding, more ducks will use an area with many small temporary marshes than will use a few large marshes.

After the ducklings hatch, the larger marshes with dense vegetation and open water become important. Now there

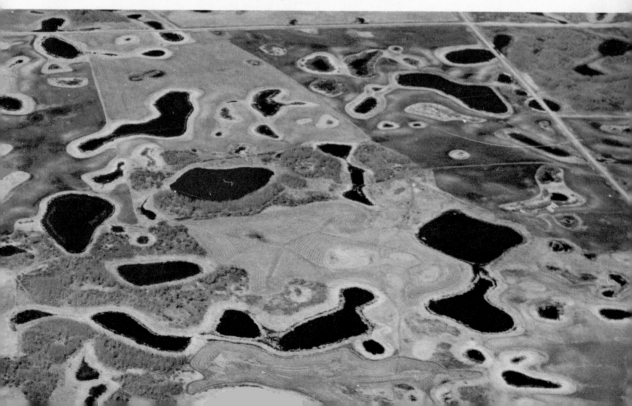

is no longer a need for isolation, but there is a need to escape predators. Young ducks, still unable to fly, can often escape from raccoons and other predators in wide expanses of marsh vegetation. In open water they can dive out of danger's way. Potholes with both open water and a ring of vegetation seem to be the ideal place for ducks to raise their newborn young.

Although representing only 10 percent of our wetlands, the pothole country produces over 50 percent of our ducks. But

In autumn this Nebraska marsh is a stopover for hundreds of thousands of waterfowl, among them pintails, canvasbacks, and Canada geese. Literally overnight thousands of birds pack the open water as wave after wave passes southward down the central flyway.

elsewhere there are productive marshes too. Every year 300,000 ducklings fly off the western marshes along the Pacific flyway. The fourteen southern states add another 700,000, and the eastern coastal marshes 200,000 during the best years. Although some are more important as the actual breeding areas, all are vital, for no young would arrive were it not for the marshes in the wintering grounds and the thousands of stepping stones in between where the weary migrants can pause for food and shelter.

179

WILLET

AVOCET

SOURIS NATIONAL WILDLIFE REFUGE

The central flyway is North America's largest area of bird migration. An important unit in the refuges along its route was formed in 1935: more than ninety thousand acres of the Souris River Valley in northwestern North Dakota were designated an area for the preservation, propagation, and protection of waterfowl, under the administration of the United States Fish and Wildlife Service. The Upper Souris Refuge is particularly popular with shore birds, many of which spend their summers there. Attracted by the small crustaceans, mollusks, worms, and insects that inhabit the shores, they run along the mudflats and sand of the marsh or wade out to probe the shallow water with their bills, looking for food.

American coots dot the water at Souris. Since these birds prefer to feed on chara and other algae, they seldom compete for food with ducks.

180

MARBLED GODWIT   DOWITCHER

Increasing productivity

Marshes and swamps are wet because the water table is at or above the surface of the land. When it rains and snows, much of the water seeps downward, building an underground reservoir, the top of which is called the water table. When wetlands are drained, this water level is lowered. The land dries out, and its wildlife disappears.

After man began to realize that drainage of wetlands does not always produce good agricultural land and destroys wildlife, he went about restoring some of them. Such restoration involves the construction of dikes and water-control structures and the planting of thousands of aquatic plants to transform these areas into productive wetlands once again.

When the water table is replenished, it is possible to increase productivity further by manipulation of the water level in marshes along the flyways, though it is unwise to change the water level in areas where ducks are nesting in great numbers. Lowering the level in midsummer, thereby exposing the muddy bottom to the air, stimulates the sprouting of thousands of seeds. You will see fields of smartweed, a plant related to buckwheat, produced in this way. Smartweeds bear tiny pink flowers, as well as tiny fruits that are relished by ducks; over 36,000 seeds were taken from the gullet of a black duck in North Carolina. Sometimes these exposed flats are hand-seeded with smartweed. At the Mattamuskeet National Wildlife Refuge in North Carolina, a field of spike rush, a tiny sedge, has been encouraged to develop by the complete drawing off of the water. In the fall this growth is flooded to attract ducks and other waterfowl that feed on the sedge.

In our tidal marshes flooding is an effective way of controlling the pestiferous salt-marsh mosquito. Since this mosquito lays its eggs on exposed wet soil, flooding at the proper time will cover its egg-laying sites and prevent breeding. The states of Delaware and New Jersey have found that this technique works in many places and eliminates the need for chemical sprays that can be extremely hazardous to wildlife and fishes.

The "re-creation" of fresh-water marshes often includes the planting of aquatic species. Among those planted are the submergent pondweeds, not only because they are easily transplanted, but also because they are a major source of food for most waterfowl. Sago pondweed has narrowly divided leaves, tender tubers, and nourishing fruit, all of which are eaten. In Montana 53 percent of the canvasback's diet consists of this plant alone.

Malheur National Wildlife Refuge is a resurrected wetland. By 1926, Malheur Lake had been drained until it was almost a desert, but a dedicated group of conservationists succeeded in impounding the water by constructing a dike and water-control system.

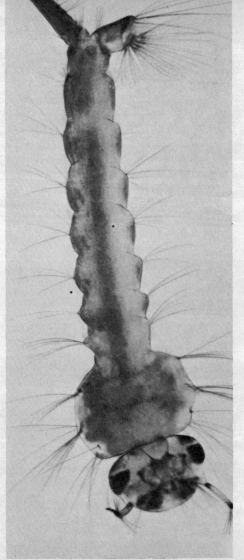

1

MAN'S FIGHT AGAINST MOSQUITOES

Most mosquitoes lay their eggs on the surface of quiet waters, where the newly hatched larvae can breathe undisturbed through breathing tubes. As a result, wetlands have always been a breeding ground for these troublesome, often deadly insects. Man's battle to control mosquitoes in marshes began as early as 1800, when oil was spread over the water surface to keep the larvae from breathing. A century later ditches were dug (*opposite*) to allow better drainage. Then during World War II, wetlands were sprayed with DDT and other insecticides. All these attempts met with some success in controlling mosquito populations, but at the same time they inflicted severe damage on marshes and their wildlife. Today an effective means of controlling mosquitoes has been found: marshes are flooded. This procedure puts the water in motion and thus decreases the larvae's chances of survival. It is even more effective against mosquitoes that lay their eggs on exposed mudflats. Probably more important, the higher water levels allow many kinds of fishes that feed on the mosquito eggs, larvae, and pupae to enter the marsh, thus controlling the numbers of adult mosquitoes and at the same time allowing an important link in the marsh food chain to continue to function.

A Culex mosquito larva hatches from the egg in a few days and then hangs head downward (1) with its breathing tube projecting through the surface film. One week later it becomes a pupa (2), and after two or three days more it emerges directly from the water as an adult (3).

3

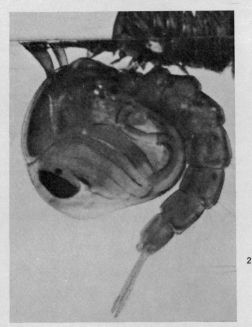

2

PLANKTON

FROG

SUNFISH

WESTERN GREBE

The animals above represent different levels of wetland food chains. Animals near the bottom of food chains have relatively small amounts of insecticide poisons in their bodies. But the concentrations accumulate to such a degree in predators higher up the food chains that they are sometimes lethal.

Decreasing productivity

Can you imagine a million fish totaling twenty to thirty tons killed within a day? Two thousand acres of a southern salt marsh had been sprayed to kill the sandfly. The killer spray was a chemical called dieldrin. Although the target was the sandfly, the shot was poorly aimed. Practically all the fish were killed, and as the fiddler crabs began to consume the dead fish, they too became the victims of the deadly poison.

In the Tule-Klamath Lake National Wildlife Refuge in California thousands of herons, gulls, grebes, and pelicans have been dying as a result of chemical pesticide poisoning for several years. Spraying is not permitted in the refuge, but the surrounding agricultural lands are being treated with pesticides, and waters from the refuge are used for irrigating these croplands. On its way through the fields the water picks up the chemical and later carries it back to the refuge. The chemical poisons accumulate in the fish and other aquatic life and are then passed on to the fish-eating herons, gulls, grebes, and pelicans.

That the poison persists and is passed on in the food chain has been strikingly shown at Clear Lake, California. Since gnats were bothering the tourists, the lake was sprayed several times from 1949 to 1957. Up until 1950 a thousand grebes had nested successfully on the marshy shores of the lake. Then for the next ten years following the first spraying no baby grebes appeared. Finally in 1962 a single baby hatched. During these years many grebes died. Five years after the spraying the DDD that was used, a chemical much less harmful to fishes than DDT, was found built up in amazing quantities in the animal life at Clear Lake. Plankton, the microscopic life of the water, accumulated 250 times the concentration of pesticide that was sprayed on the bothersome insects; frogs carried 2000 times more; and sunfish, the food of the grebes, contained over 12,000 times the concentration of pesticide originally sprayed on the gnats! Our marsh life is especially sensitive to these chemicals. They are silently taking their toll.

It is generally recognized that agricultural crops need insect control, but we must use nonpersistent chemicals or other techniques that hit the target without indiscriminately killing many other kinds of wildlife.

Along the Detroit River and its marshes twelve thousand

186

ducks, mostly canvasbacks, were killed in one month. Here the killer was oil, which is too often spilled from ships when fuel is taken on, bilge pumps are operated, or tanks are washed out. Even when oil is discharged far out at sea, it often floats into coastal areas where it endangers wildlife. When ducks and other waterfowl become covered with petroleum, their natural oils are destroyed and they are no longer waterproof. The cold water and air reach their skin, and they lose body heat faster than they can generate it. They die. Even a spot of oil no larger than a quarter can be fatal to a duck.

Detergents, modern chemical substitutes for soap, not only kill many of the marsh plants but also can destroy the oily waterproofing of birds. In Saskatchewan twenty grebes were rescued from a sewage lagoon. Only three survived. At the University of Michigan School of Natural Resources, recent experiments with catfish indicate that detergents, both the old "hard" types and the new presumably improved "soft" ones, destroy the catfishes' taste buds, which are located in their "whiskers," or barbels. Bullheads use their taste buds in finding food. When the taste buds are damaged, the fish feed less efficiently, they are weakened, and they then become more susceptible to disease.

Industrial pollution and human wastes can also destroy the life of river marshes. Many of these waste products actually overfertilize the water into which they are dumped. The result is a stimulation of aquatic life to such an extent that a severe oxygen depletion occurs. When this happens, the plants, and the animals dependent upon them, die. In the end, the river and the bordering marsh become a biological desert where no living thing can exist.

Doesn't it seem strange that there are people who work hard to restore and preserve the high productivity and beauty of our wetlands while at the same time others are destroying these resources? Some people engage in destructive activities because they are not aware of the value of these lands; others are interested only in the money they can make and have little or no concern for the effect of their actions on our wetland wildlife. Some of these wetlands, such as the Everglades and the Klamath marshes, belong to all of us, and, if for no other reason than this, they should be preserved. Moreover, all wetlands affect the well-being of every person, and they need better care than we have given them in the past.

The pure white plumage of this gannet has been soiled by a coating of oil from a waterway clogged with industrial wastes. At least 100,000 waterfowl a year are lost because of oil pollution.

Franklin's gulls, one of the sixty species of birds that breed at Bear River, nest among the reeds.

Water channels crisscross the delta where Bear River flows into Great Salt Lake. The Promontory Mountains form a rugged backdrop for this wildlife refuge.

BEAR RIVER MIGRATORY BIRD REFUGE

The white-faced ibis uses its long beak to dig crayfish, worms, and insect larvae from the mud.

In 1928, 65,000 acres of Utah wetland were set aside for what became the world's greatest game-bird sanctuary. Administered by the Fish and Wildlife Service of the United States Department of the Interior, this refuge is noted as a research center. One of its great successes is the control of botulism, or western duck sickness. This fatal duck disease has been eliminated to such an extent that there is no longer any need for the sanctuary's duck hospital where sick waterfowl were treated.

189

Wetlands regulate the water supply

In 1955 severe floods struck the northeastern United States. Many lives were lost and millions of dollars of damage was done to property. In eastern Pennsylvania, hundreds of bridges were washed out along river courses that could not contain the billions of gallons of water that surged down them. When the water subsided, two bridges, of a type that was destroyed nearly everywhere else, were left standing. Both were located just below Cranberry Bog, a natural wetland area that had been permanently preserved by a national conservation agency, the Nature Conservancy, with the aid of local Pennsylvanians. Could it be that the recent floods on the West Coast and in the Midwest are partly a result of man's destruction of marshes, swamps, and bogs?

Our wetlands are a kind of safety valve—a permanent flood-control and water-storage feature especially designed by nat-

As early as 1885, tidal marshes along the Atlantic Coast were being altered. This housing development has replaced a once-productive coastal marsh. Only in recent years have we understood the importance of these marshes as nursery grounds for many finfish and shellfish, a valuable part of our food supply. However, that is but one of the reasons for discouraging the destruction of wetlands. Much of the water captured in fresh-water marshes ultimately replenishes the water table; therefore, the loss of such wetlands can affect the water supply and also cause severe flooding.

ural forces. If the water level in a ten-acre marsh is raised just six inches, 1,500,000 gallons of water are placed in storage. In the prairie pothole country much of the water from melting snow or rainstorms is trapped in the thousands of small marshes that dot the landscape. Some of this water evaporates, but a significant amount sinks into the ground to become part of the vital underground water table from which much of our water for drinking and irrigation is drawn. The vast marshes of the southern states, the Klamath Basin in the Northwest, as well as marshes everywhere perform essentially the same role. And it is an important one.

Just as important, water running off the land is slowed to a rate that can be handled by existing river channels. Flooding occurs when run-off increases to a point where the channels of streams and rivers can no longer contain the water flowing into them. The filling of our marshes and swamps has only intensified the problem of flood control.

Not all the hides brought in to the fur companies were trapped by the mountain men; many were purchased by trade with the Indians. Even as early as 1620, the year the Pilgrims landed in Massachusetts, there were at least a hundred fur traders bartering for pelts with the Indians along the Chesapeake Bay of Virginia and Maryland.

The fur trade

American history was greatly influenced from the outset by the beaver trade. For four centuries the felt manufactured from beaver fur was used universally to give the civilized world its hats. The quest for beaver was conducted exhaustively. The first explorers, looking for a shorter route to the East, found a vast continent teeming with rivers, lakes, marshes, and bays. Every coast they visited, from Massachusetts Bay to Florida, every river they sailed up, every wetland they explored was swarming with beaver, otter, muskrat, and mink.

Royal charters granting hunting territories were given by the French and English kings, great mercantile companies were established, and settlements were planted in the New World from Quebec and Massachusetts Bay to Georgia. On the Hudson River the Dutch founded a fur depot where New York City now stands. Its outpost, which they called Beberwyck ("Beavertown"), was renamed Albany when the English seized the whole region.

From Quebec, the great bastion of the French beaver trade, companies of armed men paddled through the Great Lakes and down the Mississippi to the Gulf. Along the way they dealt with the Indians, signed monopolistic trade agreements, and arranged systematic exploration of the fur resources in their wetlands. The English colonists on the Atlantic Coast made similar arrangements with the tribes. Company agents in Charleston, South Carolina, were soon receiving furs taken in the flood-plain wetlands of the Mississippi and from the marshes of Georgia and Alabama and the mountain lakes of Kentucky.

The deeper interior of the continent was similarly opened up by traders looking for new sources of beaver. Water routes were explored and mapped, and permanent trading posts were established; these were fortified and grew into cities. Pittsburgh, Detroit, Chicago, St. Louis, and New Orleans were a few such trading posts. Rivalry was incessant

The hardiest of the fur trappers and traders were the mountain men who hunted alone in the wilderness beyond the frontier. To survive, they learned to hunt, live, and eat like the Indians. Even their clothing — leather shirt, breeches, and moccasins — were Indian-made. Always armed, the mountain man usually stuck his skinning knife, bullet pouch, and hatchet in his leather belt; and he never went far without a powerful rifle.

between the French and the English, and their struggle for possession of the beaver market culminated in the French and Indian War. When the war ended, the English were the masters of the fur trade on most of the continent.

The profit derived from furs constituted half the annual income of Canada as late as a century ago. For more than fifty years after the American Revolution, it was hardly less significant in the United States. It was beaver that lured venturesome men westward through the Rockies and into the Southwest.

By then these explorers were no longer trading with the Indians; they were trapping the animals themselves. Their long canoe files proceeded down rivers and overgrown waterways, bearing well-equipped companies of fifty to a hundred men. They worked over the marshes systematically, using steel traps. These experienced woodsmen learned to read "sign," the traces of animal presence in a wild place. From tracks, bones, feathers, gnawed bark, fish scales, drag marks, or the flattening of sedges, they could deduce entire histories. They developed great skill in telling where the hunting would be good.

A buyer inspects fox, mink, squirrel, muskrat, and Persian lamb pelts in a fur merchant's storeroom. The merchant is a middleman, obtaining his goods from trappers and fur farmers at one of the large auction centers—New York, Chicago, St. Louis, or Montreal —and then selling them to furriers.

The fur trade today

About 1845, beaver fur was trapped out and lost its preeminence as the material for felt hats. The fur of the nutria, a mammal introduced from South America, began to be used more widely. The beaver populations had a respite. Today they are again trapped extensively in over thirty states, and nearly 200,000 pelts are taken annually.

But muskrats have now become the most important fur bearers in the United States. Nearly 5,000,000 of them are trapped in most years. Iowa, Minnesota, Ohio, Pennsylvania, and Wisconsin are the most productive states; each contributes half a million pelts or more every year. Last year a muskrat pelt brought 80 cents to $1.45 on the fur market. Today many of our wetland management programs are aimed at increasing the productivity of this valuable mammal.

As you have learned, muskrats thrive best when they can forage for food in open water of a suitable depth throughout the year. Unlike beavers they store nothing. Sound marsh management programs for muskrats frequently employ a practice called blind ditching. The ditches are not for drainage, but simply are dredged out to provide areas of deep water that will not freeze to the bottom during the winter. Moreover, their soft earth banks are important places for muskrats to build their lodges.

From the wetlands trappers have derived an income from muskrat pelts totaling as much as five million dollars or more a year. In addition, over a million raccoons were caught in marshes and swamps, over 400,000 nutrias, 300,000 minks, 100,000 beavers, and 12,000 otters.

Minks are much less numerous and more difficult to catch than the others, and so the pelts are relatively costly. They are greatly prized because of their silky quality. But the mink is a resourceful, intelligent species, and a century of intensive hunting and trapping has not greatly reduced its numbers.

Wetlands for the future

Cattails rustle in the cold winter wind as a flock of mallard ducks huddles in the protection of the winter stubble. A clapper rail slips silently through the cordgrass on a tidal marsh. Along the edge of a watercourse in the Everglades an anhinga spreads its wings to dry, and in the swampy waters below an alligator swims through a canal choked with water lettuce. Far north of cypress country a snow-shoe hare bounds off across the yielding sphagnum mat of a spruce bog.

You have journeyed from one kind of wetland to another, from a bog to a swamp, a fresh-water marsh to a salt-water marsh. You have seen how life in a watery environment has led to the evolution of myriad adaptations: floating and emergent plants, webbed feet, life cycles partly in water and partly on land, long bills and long legs for feeding in shallows—a marvelous assemblage of adaptations uniquely fitting plants and animals to life in the wetlands.

Although some people are beginning to realize the value of our marshes, swamps, and bogs, we have only begun. The famous Everglades area of Florida, the ancestral home of vast numbers of subtropical wading birds—ibises, egrets, and spoonbills—is threatened by schemes for the artificial control of natural water supplies for agricultural and land-reclamation purposes. Because of ditching, filling, and draining, you have to be lucky to see black ducks nesting in many of our coastal salt marshes.

Wetlands have always produced a rich crop of plants and animals and have provided the food necessary for

The nutria, or coypu, is a newcomer to North America. These large rodents were imported from South America in 1899 and released to breed in southern marshes. In contrast to the muskrat, nutrias only "top" aquatic plants rather than eat them out, roots and all. Their reddish-brown fur is becoming increasingly popular in the world of fashion and is often used as a substitute for beaver.

Today, more often than not, the
scene along both the Pacific
and Atlantic Coasts is one of
prosperous industry. The
progress of the twentieth
century has left the old natural
land a poorer place, as a
growing demand for material
products has called for an
increase in factories. Yet
ironically, some of the material
things—oysters, game fish,
muskrat coats, and nutria
bedspreads—depend in a very
real way upon the existence of
the marshes that are being
replaced by resort developments,
factories, and garbage dumps.
A few precious acres have been
set aside as national refuges
and will remain inviolate.
But for the rest of our rapidly
dwindling coastal wetlands the
future is uncertain.

coastal shellfish. They have nurtured thousands upon thousands of ducks and geese, which fed the first settlers in Colonial America. Later they furnished the beavers and muskrats that helped shape a national economy.

But even before the arrival of the white man, marshes, swamps, and bogs served the original Americans. Around the Great Lakes many tribes found nourishment from the vast acres of wild rice that they harvested. Blackfoot Indians were always great trappers of the mink, which they used for food and clothing. They marveled at the mink's intelligence and fighting spirit when it was faced with an enemy.

The forms of life inhabiting these wetlands have developed over millions of years, yet man can destroy them all in less time than he realizes. They should not die, for their worth to man is greater than he has yet realized. Feeding oceans, stocking flyways, storing water, preventing floods, providing recreation and a new adventure in living for those who explore them—these are but a few of the roles our valuable wetlands play. There is still time to save these unique national treasures.

The future of wetlands is entrusted to the present generation. For centuries to come, youngsters like these should be able to explore a marsh, to see its plants and animals, and to sense the millions of years involved in its making.

Appendix

Wetland Areas in the National Park System

A small but significant portion of the vast wetland resources of the United States lies within the boundaries of the national parks and seashores.

The northern bogs and muskegs of Maine and Alaska, with their dazzling variety of wildlife and vegetation, are on view in national parks. Tidal marshes are open to visitors to the national seashores on the East and West Coasts. North, south, east, and west, almost every conceivable type of wetland, in varying stages of development, can be explored. Areas such as Grand Teton National Park in Wyoming boast colorful and fascinating arrays of plant and animal life.

The national parks were created as a living museum to preserve America's natural heritage. National seashores, on the other hand, are mainly recreational areas. But visitors to both may enjoy such activities as picnicking and camping while they observe and learn about America's forests, lakes, and mountains, as well as her wetlands.

In many of the parks and seashores National Park Service naturalists conduct guided tours. Their work is supplemented by self-guiding nature trails, illustrated evening campfire programs, museum exhibits, and publications.

A few of the most important wetland areas included in the national parks and seashores are described here.

Acadia National Park (Maine)

The Big Heath, one of the most striking examples of a black spruce bog in the National Park System, is on the western side of this park, near the Seawall Campground, in the southwestern portion of Mount Desert Island. Besides the majestic stand of encircling spruce, the bog displays an abundance of sphagnum moss and carnivorous pitcher plants. Other bogs, in varying stages of development, are scattered over the island. Additional wetlands in the park include Newmill Meadow, on Duck Brook; Gilmore Meadow; and Great Meadow, on Ocean Drive near Sieur de Monts Spring. A well-kept system of foot trails leads visitors through the park. The Big Heath itself is accessible by road. Other bogs are just outside the park, beyond the northern boundary. A salt-water marsh lies next to Thomas Bay.

Assateague National Seashore (Maryland, Virginia)

Extensive salt-water marshes run along the bay side of this slender barrier island on the Atlantic Coast. Ducks, geese, and swans winter in the Chincoteague National Wildlife Refuge, which occupies the southernmost part of Assateague. See pages 54 and 55.

Cape Cod National Seashore (Massachusetts)

Nauset Marsh attracts thousands of visitors, who take the guided tours conducted by rangers of the National Park Service on this narrow sand bar in the Atlantic Ocean. Algae, sponges, snails, clams, crabs, and worms find shelter on Nauset, with its great tidal flats and lush grass. There are fine examples of swamps and bogs, as well as other marshes, here. Eastham Bog glistens with swamp azaleas and highbush blueberry. Red Maple Swamp has an abundance of bird life and dense growths of blueberry, sweet pepper bush, and green brier. Atlantic White-cedar Swamp, east of Drummer Cove near Wellfleet, is the only cedar swamp on the seashore. The Herring River Marshes are nearby. At the north end of the Cape hook, near Truro, are Pamet Valley Cattail Marsh and Featherbed Swamp. Both fresh- and salt-water marshes as well as bogs are found near Provincetown.

ATLANTIC WHITE CEDAR

Cape Hatteras National Seashore (North Carolina)

Salt- and fresh-water marshes are extensive on this chain of barrier islands extending seventy miles along the coast. Tidal salt-water marshes fringe Roanoke and Pamlico Sounds along the west side of Hatteras and Bodie Islands. Smooth cordgrass, salt-meadow cordgrass, black rush, salt grass, marsh alder, and silverling are much in evidence. In the area of the Cape Hatteras Lighthouse, near Buxton, are fresh-water ponds edged with cattails and saw-grass marshes. A self-guiding nature trail describes the wetlands at Buxton Woods. Fresh-water marshes neighbor the Bodie Island Lighthouse. The Pea Island National Wildlife Refuge is on the fresh-water marshes on the north end of the Hatteras Island. Here thousands of greater snow geese, many species of ducks, and whistling swans spend their winters.

Cape Lookout National Seashore (North Carolina)

Tidal marshes border Core Sound for almost the entire length of this section of the Outer Banks. One of the newest of the national seashores, Cape Lookout extends fifty-eight miles over a series of islands from Ocracoke Inlet to Beaufort Inlet.

Everglades National Park (Florida)

The Everglades marshes and mangrove swamps lie in this water park of nearly two thousand square miles on the Florida peninsula. The marshes, dominated by saw grass, are dotted by small hardwood tree islands called hammocks. The swamps occupy a broad belt that extends inland to meet the long, curving river of saw grass. Some of the red, black, and white mangroves reach heights of forty feet and more and are probably the tallest of their kind in the world. Visitors to the marshes and swamps travel along self-guided trails elevated above the wet ground. Signs and recorded talks describe the highlights of the areas. Regularly scheduled boat tours are conducted through the mangrove areas. The bird life of the Everglades is one of the wonders of the National Park System. The rare roseate spoonbill, the wood and white ibises, and seven species of herons make their home here. See pages 14 to 17.

Glacier National Park (Montana)

Wet meadow marshes occupy many of the depressions in this park in the Rocky Mountains. Probably the most interesting are the meadows along Camas Creek and McGee Creek on the western side of the park, and in Waterton Valley and Belly River on the eastern side. The sphagnum and sedge of a bog are gradually encroaching upon tiny Johns Lake. Guided nature walks are conducted to this bog from Lake McDonald Lodge.

Grand Teton National Park (Wyoming)

There are extensive marshes and bogs in this park six miles south of Yellowstone National Park. The full range of wetland animal and plant life is found here. Otters, minks, moose, snowshoe rabbits, and beavers flourish. Above the marshes the osprey hovers. Waterfowl include the rare trumpeter swan (from nearby Red Rock Lakes) and mergansers, baldpates, and mallards. Black spruce, larch, and a variety of shrubs neighbor near smaller plants, such as gentian, marsh bluebell, lupine, meadowsweet, goldenrod, and bluegrass. An excellent trail system covers much of the park.

OTTER

Isle Royale National Park (Michigan)

Geologists believe this island was submerged for many years after the glaciers receded, emerging only after the level of what is now Lake Superior had fallen. The 210 square miles of wilderness include lakes, ponds, swamps, and bogs occupying the depressions between the parallel ridges extending from one end of the island to the other. Two types of bogs, in varying stages of development, may be seen: the sphagnous, with sedge mat, sphagnum moss, black spruce, little or no drainage, and relatively deep basins; and the cyperaceous, with less sphagnum, shallower basins, active outlets, and white cedar or arborvitae as the primary tree species. A large black spruce swamp, the Big Siskiwit Swamp at the head of Siskiwit Bay, is an example of the kind of swamp that occurs on the island where the soil is acid. A northern white cedar swamp on the northeastern shore of Lake Richie is typical of the kind that occurs in slightly alkaline soil. Moose roam the island.

Mount McKinley National Park (Alaska)

Muskegs are plentiful in these three thousand square miles dominated by the highest peak on the North American continent. Muldrow Glacier, thirty miles long, descends from the northeastern flank of the mountain. Black spruce, Labrador tea, sedges, and lichens characterize the muskegs, which overlay and insulate the permafrost subsoil in scattered depressions. Moose, red foxes, beavers, minks, and snowshoe hares are frequently seen in the vicinity of the muskegs.

MOOSE

National Capital Parks (District of Columbia, Virginia, Maryland)

In the area around the nation's capital, wetlands of interest include the following: a well-developed cattail marsh and a silver maple swamp on Theodore Roosevelt Island in the Potomac River, which can be reached by ferry or on foot over a causeway; Dyke Marsh, an extensive cattail marsh along the Mount Vernon Parkway near Belle Haven; several swamps in the vicinity of Kenilworth Aquatic Gardens and along the Anacostia River, which are covered in guided nature walks open to the public during the summer months.

Shenandoah National Park (Virginia)

A swamp near the Big Meadows Campground in this section of the Blue Ridge Mountains is one of the rare examples of wetlands along the crest of the mountains. A self-guiding nature trail winds along the edge of the swamp.

Yellowstone National Park (Idaho, Montana, Wyoming)

The ideal way to explore the marshes of this park is by foot or on horseback. The marshlands along the Upper Yellowstone River and along Bechler River in the southern part of the park are accessible only by trail. The marshes along Slough and Pelican Creeks are nearer to roadways. Beavers and moose are plentiful here.

Yosemite National Park (California)

Swamp Lake, near Lake Eleanor, and the floating islands are the most impressive of the wetlands in this park in the Sierra Nevada Mountains. At all elevations the visitor will find meadow marshes along the streams and lakes. Eighty-five percent of the park is wilderness.

206

National Wildlife Refuges

Some of the finest scenic wetlands in the United States are found in the national wildlife refuges, a system of nearly thirty million acres in three hundred separate areas administered by the Fish and Wildlife Service of the Department of the Interior. These carefully protected lands and waters are among the last remaining places where a visitor can see the great congregations of waterfowl and other birds as they migrate over the national flyways. Most of the refuges have been established for the protection of waterfowl. Some are principally for migratory birds other than ducks and geese, such as pelicans, herons, egrets, ibises, and spoonbills. Others are sanctuaries for endangered species; for example, the Aransas Refuge in Texas is the home of the whooping crane, and the rare trumpeter swan nests at Red Rock Lakes in Montana. All the refuges are richly varied habitats for mammals, amphibians, reptiles, insects, and plants.

Each year more than twelve million visitors pass through the refuges to look, to photograph, to picnic, and, where permitted, to hunt and fish. The highlights of some of the refuges are discussed below. Others are located on the map on pages 210 and 211.

Aransas National Wildlife Refuge (Texas)

The rarest bird on the North American continent, the whooping crane, makes its winter home in this seventy-four-square-mile area on the Gulf coast. The five-foot white birds, with a wingspread of seven feet or more and a clear, buglelike call, begin arriving in October and stay till late March. The refuge is composed of lakes, ponds, wooded areas, marshy swales, and coastal marshes on the fringes. The southeastern strip of flatland, cut by tidal sloughs and estuaries, is the favorite feeding ground of the whoopers as well as thousands of ducks and geese. Among the nearly three hundred species of birds are the rare roseate spoonbill, with its bright pink plumage; the gray sandhill crane; the snowy, reddish, and American egrets; and the great blue and Louisiana herons. In the wooded swamplands are white-tailed deer, the nine-banded armadillo, and the peccary, or javelina, the only native wild pig in North America.

TRUMPETER SWAN

Bear River Migratory Bird Refuge (Utah)

In autumn hundreds of thousands of waterfowl and shore birds fly into this refuge on the delta marshes of the Great Salt Lake, between the Pacific and central flyways. The big and graceful white whistling swan, Canada and lesser snow geese, and ducks —widgeons, canvasbacks, scaups, and the rarer scoters and hooded mergansers—are on view. Shore birds include many varieties of sandpipers, snipes, long-billed curlews, and marbled godwits. The refuge is one of the few remaining wintering grounds for white pelicans, whose wings hum in flight as the wind passes through. In spring, on the return trip north, come Canada geese, pintails, teals, scaups, and four species of grebes —western, eared, horned, and pied-billed. The black-necked stilt, with its black and white plumage, and its close relative, the avocet, with up-curved bill, are among the waders.

BLACK DUCK

Bitter Lake National Wildlife Refuge (New Mexico)

Bitter Lake is a salty alkaline body of water near the meandering Pecos River. Nine other lakes on this refuge are impounded by dikes in the lowland peat bogs and marshes that were the prehistoric channels of the river. On a tract overlooking the lakes, land is irrigated to grow grain and green browse for waterfowl. The refuge, the only water area of its kind along the Pecos River, is a stopping place in the fall migration of more than twenty species of waterfowl, particularly mallards, American widgeons, pintails, and canvasbacks. The lesser sandhill crane arrives from its Arctic nesting grounds at the end of September and leaves in early March. In the fall, too, white pelicans and double-crested cormorants are common. Among the shore birds are black-necked stilts, snowy plovers, and American avocets. In some areas hunting is permitted, and there are shaded picnic tables overlooking the lakes.

Blackwater National Wildlife Refuge (Maryland)

Dark loblolly pines rise on the borders of the marshes in this flourishing waterfowl area on the eastern shore of Chesapeake Bay. It contains ten thousand acres of marsh, timbered swamps, brush, and fresh-water ponds. A population of muskrats, held down to suitable levels by trapping, helps maintain the balance and well-being of the waterfowl habitat. At the peak of the fall migration more than 150,000 ducks—mallards, black ducks, pintails, green-winged and blue-winged teals, American widgeons, and wood ducks—stop here. Over 230 species of birds have been spotted. In addition to waterfowl there are marsh birds, shore birds, loons, and grebes. The twittering long-billed marsh wren builds its nest in the cattail marshes. Sightseers will find the dike trails useful, especially in early morning, when waterfowl are most active.

Cape Romain National Wildlife Refuge (South Carolina)

Islands dot the salt marshes of this fifteen-mile segment of the Atlantic Coast. The most popular is Bulls Island, a barrier reef that tapers off on its western side to the great tidal marshes. In the winter the island is taken over by widgeons, pintails, black ducks, scaups, ring-necked ducks, gadwalls, redheads, mergansers, and buffleheads. Loons, horned grebes, and green, great blue, and Louisiana herons abound. Coots and common gallinules feed along the margins of ponds. Alligators are a common sight along the water's edge. In the summer, when the waterfowl are gone, wood ibises and common and snowy egrets may be seen. The clapper rail is a year-round resident. The mudflats and sand bars between the islands are covered by shore birds—dowitchers, whimbrels, marbled godwits, sandpipers, black-bellied and ringed plovers, ruddy turnstones, and the oystercatchers, for whom the refuge was established.

Kenai National Moose Range (Alaska)

The western two-thirds of this subarctic range are lowlands—rolling hills, low, ridges, and muskegs interspersed with more than twelve hundred lakes. The area was set aside principally for the moose, the fifteen-hundred-pound monarch of the deer family, whose long legs carry it through the deep snows in winter and the marshes in summer. The lowlands are covered with aspen, willow, birch, and, in the bogs, black spruce. Fragrant Labrador tea and small cranberry grow in profusion. During spring and fall migrations thousands of waterfowl congregate, particularly in Moose River near the Sterling Highway, at the mouth of Skilak Lake, and on the Chickaloon Flats. In the summer mallards, goldeneyes, teals, scaups, and harlequin ducks nest on the lake shores, while lesser Canada, emperor, and white-fronted geese and brants pass through. The trumpeter swan, now nearly extinct, sounds its trumpetlike call from one lake to another. Minks, beavers, muskrats, and otters are among the mammal life.

Lower Souris and Upper Souris National Wildlife Refuges (North Dakota)

The two refuges on the Souris River are narrow areas extending along the valleys. In summer Lower Souris is the breeding place of the grain-feeding ducks—mallards and pintails—and some dozen other species, as well as the central flyway stopover for grebes—red-necked, pied-billed, eared, horned, and western. The marshes are also home to Franklin's gulls, double-crested cormo-

SMALL CRANBERRY

CANADA

FLATTERY ROCKS
QUILLAYUTE NEEDLES
COPALIS
WILLAPA
CAPE MEARES
THREE ARCH ROCKS

JONES I.
SAN JUAN
MATIA L.
SMITH I.
DUNGENESS
WASH.
TURNBULL
COLUMBIA
MCNARY
COLD SPRINGS
MCKAY CREEK

CREEDMAN COULEE
L. THIBADEAU
PABLO
PISHKUN
NINE PIPE
WILLOW CR.
BENTON L.

BLACK COULEE
HEWITT L.
BOWDOIN
MEDICINE L.
WAR HORSE
L. MASON
HAILSTONE
HALFBREED L.
LAMESTEER

LOWER S
UPPER SOURIS
N. D
MCLEAN
WHITE L.
A
JOHNS
ARRO
PRETTY R
SAND

OREGON

KLAMATH FOREST
UPPER KLAMATH
OREGON IS.
TULE L.
LOWER KLAMATH
CLEAR L.
MODOC

MALHEUR

RED ROCK LAKES
IDAHO
DEER FLAT
CAMAS

MINIDOKA
LOCOMOTIVE SPRINGS
BEAR R.

MONT.

WYOMING
PATHFINDER

BAMFORTH
HUTTON L.

BEAR BUTTE W
BELLE FOURCHE
S. DAKO
LACREEK
L. AND
VALE
NORTH PLATTE
CRESCENT L.
NEBRAS

SACRAMENTO
DELEVAN
COLUSA
FARALLON

SUTTER
ANAHO L.
FALLON
STILLWATER

RUBY L.

FISH SPRINGS
OURAY

CALIF.
MERCED
PIXLEY
KERN

NEVADA

UTAH

COLORADO

MONTE VISTA

BURFORD L.

KANS
QUIVIRA

SALT PLAIN
W

HAVASU L.
ARIZONA

SALTON SEA
IMPERIAL

NEW MEXICO
BOSQUE DEL APACHE

BITTER L.

BUFFALO LAKES
MULESHOE
OK

CHAMISSO

ALASKA

BERING SEA
HAZEN BAY
TUXEDNI

PRIBILOF

BOGOSLOF
SEMIDI

ST. LAZARIA
HAZY IS.
FORRESTER I.

HAWAII

MEXICO

TEXA

0 200 400 MILES

JOHNSTON I.
0 200 400 MILES

LAGUNA ATASC
SANTA AN.

NATIONAL WILDLIFE REFUGES
FOR MIGRATORY BIRDS

AGASSIZ

TAMARAC

RICE L.

MILLE LACS

MINN.

TREMPEALEAU

NION SLOUGH

IOWA

SOTO

SQUAW CREEK

SWAN L.

MISSOURI

MINGO

MINGO

MAN

ARK.

HOLLA BEND

WHITE R.

WAPANOCCA

BIG L.

MISS.

CATAHOULA

YAZOO

DAVIS I.

NOXUBEE

LA.

SABINE

LACASSINE

SHELL KEYS

E. TIMBALIER I.

HORN I.

PETIT BOIS

BRETON

DELTA

HURON

SENEY

GREEN BAY

GRAVEL L.

WISC.

NECEDAH

HORICON

MICH IS.

SHIAWASSEE

MICH.

WYANDOTTE

W. SISTER

OTTAWA

IND.

ILL.

CHAUTAUQUA

CRAB ORCHARD

KENTUCKY

KENTUCKY WOODLANDS

REELFOOT

L. ISOM

CROSS CREEKS

TENNESSEE

TENN.

WHEELER

PIEDMONT

ALA.

GEORGIA

OAK ORCHARD

L. ST. CLAIR

OHIO

ERIE

PA.

W.VA.

VIRGINIA

PRESQUILE

BACK BAY

MACKAY I.

MATTAMUSKEET

SWANQUARTER

N.C.

PEA I.

CAROLINA SANDHILLS

SANTEE

S.C.

CAPE ROMAIN

SAVANNAH

BLACKBEARD I.

OKEFENOKEE

TYBEE

HARRIS NECK

WOLF I.

ST. MARKS

BREVARD

CEDAR KEYS

CHASSAHOWITZKA

ANCLOTE

PINELLAS

ISLAND BAY

SANIBEL

FLA.

PELICAN I.

PASSAGE KEY

LOXAHATCHEE

GREAT WHITE HERON

MISSISQUOI

MONTEZUMA

N.Y.

GREAT SWAMP

N.J.

KILLCOHOOK

BRIGANTINE

SUSQUEHANNA

EASTERN NECK

BLACKWATER

MARTIN

PRESQUILE

BOMBAY HOOK

DEL.

MD.

CHINCOTEAGUE

MAINE

MOOSEHORN

VT.

N.H.

PARKER R.

GREAT MEADOWS

MASS.

CONN.

R.I.

MONOMOY

MORTON

WERTHEIM

● Migratory Bird Refuge

0 50 100 150 MILES

rants, white pelicans, coots, common terns, great blue and black-crowned night herons, avocets, and yellow-headed and red-winged blackbirds. Beavers, muskrats, minks, and raccoons move stealthily through the water and mudflats. On the higher ground are white-tailed deer, rabbits, and red foxes. Upper Souris is the nesting place of whistling swans and geese. On Lake Darling, largest of the six bodies of water on the refuge, visitors may see two colonies of western grebes, a large rookery of double-crested cormorants and great blue herons, and hundreds of white pelicans. The willet, avocet, marbled godwit, and Wilson's phalarope are among the shore birds. Large populations of beavers, muskrats, minks, and raccoons roam the area.

BLACK BEAR

Loxahatchee National Wildlife Refuge (Florida)

This 228-square-mile wilderness area is the southernmost wintering ground in the continental United States for the waterfowl of the Atlantic flyway. Teal, pintails, lesser scaups, ringnecks, redheads, mallards, and wood ducks stop here. The limpkin, a brown and white-spotted wading bird the size of a bittern, with a downward-curving bill, is one of the inhabitants of the refuge. The largest bird here is the gray Florida crane. Others include the white ibis and the American and snowy egrets. The water areas are dotted with small islands of dense vegetation. Shrubs and trees include white bay, myrtle, strangler fig, and moss-draped cypress. The shallow waters around the island abound with water lilies and pickerelweed. To aid the waterfowl, the Fish and Wildlife Service has planted Egyptian wheat, sorghum, millet, and rice. Bobcats, otters, raccoons, alligators, and white-tailed deer may be seen here. Loxahatchee is one of the last remaining sanctuaries for the nearly extinct everglade kite.

Malheur National Wildlife Refuge (Oregon)

Among the more than two hundred species of birds that stop at this refuge is the sandhill crane, whose mating dance may be seen in the spring. In the late afternoons these big forty-inch-tall birds, with heads thrown high, begin to hop slowly, criss-crossing each other's path, and increase their speed until only a blur can be seen. A dozen species of ducks, whistling swans, Canada geese, handsome black terns and Forster's terns, common and snowy egrets, and white-faced ibises nest in the marsh. Avocets stalk the shallows; the cries of willets, killdeer, and Wilson's phalaropes are heard; and the startling long-billed curlews move through the salt-grass flats. Along the great valley of the Blitzen River, where volcanic action has deposited grotesque rock formations and stately basaltic columns, minks and muskrats thrive and beavers build their dams and houses.

212

Moosehorn National Wildlife Refuge (Maine)

This 22,565-acre refuge on the eastern coast of Maine was established in 1937 to protect the habitat of waterfowl, spruce, and other life. One part of the area borders the tidal waters of Cobscook Bay, a branch of Passamaquoddy Bay. Sea lavender grows along the tidal margins, and tides there rise and fall about twenty feet. Spruce, larch, and cedars are among the trees that predominate. At Moosehorn itself—the larger of the two units comprising the refuge—the mammal population includes white-tailed deer, black bears, skunks, raccoons, red foxes, beavers, minks, otters, weasels, muskrats, and, sometimes, moose. Among the ducks are ringnecks, blacks, American goldeneyes, oldsquaws, and mergansers. Canada geese and Atlantic brants stop over. Circling above are the predators—sharp-shinned, red-tailed, broad-winged, and marsh hawks.

Okefenokee National Wildlife Refuge (Georgia)

The refuge occupies about four-fifths of Okefenokee Swamp, one of the oldest and most primitive swamps in America. The swamp is a great morass of mossy cypresses and watery prairies. The name Okefenokee is the white man's version of the Indian words for "Land of the Trembling Earth." Visitors stamping on the thick and often unstable peat floor can cause nearby trees to tremble. The big cypresses are not in contact with the ground but are rooted in the upper crust of the peat bed. Alligators, bears, raccoons, muskrats, and otters live in the swamp, along with many species of water birds, including the egret, sandhill crane, ibis, wood duck, and anhinga, or snakebird, which can swim submerged with only its head and snakelike neck above the surface. In the sixty thousand acres of open marsh are white and yellow water lilies, neverwet, pickerelweed, floating hearts, swamp marigold, and the carnivorous bladderwort and pitcher plant.

VIRGINIA RAIL

Upper Mississippi River Wildlife Refuge (Illinois, Iowa, Minnesota, Wisconsin)

The boundaries of this nearly 200,000-acre area of marshes, wooded islands, and waters are the longest of any inland Federal refuge. They extend nearly three hundred miles in four states, along river bottoms from Wabasha, Minnesota, to Rock Island, Illinois, and encompass a wide variation in life zones and climate. A system of dams has created marshlands, channels, sloughs, and open lakes over the bottom lands. The river valley is a major migratory route for whistling swans, herons, egrets, bitterns, rails, and ducks. Diving ducks include lesser scaups, ring-necked ducks, redheads, canvasbacks, buffleheads, and ruddies. Among the surface feeders are mallards, widgeons, gadwalls, and teals. Brilliantly marked wood ducks feed in the sloughs and shallows and nest in the trees of the islands and bluffs. Fur-bearing mammals include the muskrat, mink, beaver, otter, raccoon, skunk, weasel, fox, and a small number of nutria, which have appeared in recent years.

Red Rock Lakes National Wildlife Refuge (Montana)

The rare trumpeter swan, the largest North American waterfowl, lives in this refuge, which is just west of Yellowstone National Park and seven thousand feet above sea level. The principal homes of the swans are shallow lakes and marshes and the islands of bulrushes and other sedges. Shiras' moose browse on the leaves of the willow tree. Bears, beavers, and minks are nearby. There are few muskrats, but they are important: the trumpeters nest in occupied or abandoned muskrat houses separated from open water by sedges or bulrushes. The water ouzel, a gray bird, plies the streams in search of the larvae of caddis flies and other aquatic insects. Virginia and sora rails are found among the tules, and on the muddy lake borders are great blue herons and avocets. Among the other neighbors of the trumpeter swan are the marsh wren, black-crowned coot, night heron, American bittern, black tern, and sandhill crane.

Reelfoot National Wildlife Refuge (Tennessee)

BLACK CRAPPIE

Scattered giant cypresses grow during late spring and summer in Reelfoot Lake. The water is paved with yellow pond lily pads, and the border is an almost impenetrable swamp forest giving way to mats and man-high grasses in shallow waters. In winter, mallards, American widgeons, gadwalls, and ring-necked and wood ducks take refuge, feeding mainly on pondweeds, naiads, and duckweeds. Egrets, herons, cormorants, gallinules, and gulls are among the more than two hundred species of birds. The anhinga, or snakebird, perches on the cypress trees or swims with only its head and long neck showing above the surface. In the winter evenings millions of grackles, red-winged blackbirds, cowbirds, and starlings swarm into the lake to roost in the cutgrass. Muskrats and raccoons are common. Sport fishing is permitted. Crappie, bream, carp, and spoonbill catfish are abundant. The big snapping turtle, with its algae-covered carapace, moves through the waters of the lake.

Seney National Wildlife Refuge (Michigan)

The rare sandhill cranes perform their wild courtship dance on this refuge of diked and sedgy marshes and river bottoms near the Canadian border. In late summer and early autumn, when the area is crowded with waterfowl, the cranes come out of their secluded nesting places to feed along the marshy lowlands. Common loons, Wilson's snipes, and yellow rails inhabit the marshy areas. Among the waterfowl Canada geese and ducks such as blue-winged teals, widgeons, mallards, black ducks, ring-necked ducks, and American and hooded mergansers are prominent. The wood duck is here in the hollow trees of the swamps and borders of streams and ponds. Mammal life includes white-tailed deer, beavers, muskrats, minks, foxes, raccoons, and, occasionally, black bears and moose. Auto caravan tours are conducted on summer evenings at six o'clock.

Tule Lake, Lower Klamath, and Sacramento National Wildlife Refuges (California)

Three to four million waterfowl of some 250 different species flock each autumn to the Klamath Basin in northern California and southern Oregon and to nearby Tule Lake, way stations on the Pacific flyway. Their 60,000 acres are refuge for 600,000 geese, including the white-fronted, Canada, lesser Canada, and the rare little Ross' goose, a species of the lesser snow goose, no bigger than a mallard. There are whistling swans, white pelicans, and a wide variety of ducks, including the vanishing redhead, which needs large marsh areas for survival, the cinnamon teal, the gadwall, and the ruddy. Among the waders are sandhill cranes, white-faced glossy ibises, herons, and bitterns. At sundown the short-eared owl skims and swoops over the marshes. Upper Klamath, on the Oregon side of the basin, is a unique ecological environment, with its vast tule marsh, drowned stream channels, willow-lined banks, and coniferous forests on steep mountain slopes. Some 150 miles to the south is the Sacramento Refuge, principal wintering ground of the Ross' goose and autumn home for more than a million pintails.

WATER TUPELO

White River National Wildlife Refuge (Arkansas)

Spring rains flood the forest and the river bottoms of this narrow strip of refuge. The White River, a Mississippi tributary, winds and twists through the area. Crescent-shaped lakes bordered by cypress trees dot the forest. The trees include water tupelo, black and sweet gum, and swamp privet. During the spring flooding, visitors explore the refuge by boat. In drier seasons, trails and dirt roads may be used. King rails, coots, short-billed marsh wrens, and the waders—egrets and herons—live in or pass through the area. More than a half million mallards ply the lakes in early January, joining blue-winged and green-winged teals, pintails, and widgeons. Honking Canada geese also stop over. The wood duck and the hooded merganser are the only waterfowl that nest on the refuge. Mammals to be seen include the black bear, white-tailed deer, raccoon, mink, beaver, and otter.

Wetland Frogs and Toads

A fresh-water wetland in spring can be a noisy place. One by one the different species of frogs and toads begin to sing their mating calls—snorting, trilling, and whistling to attract a mate. First the spring peepers and wood frogs sing out. By May, the other frogs join the chorus, producing a deafening din. Then gradually the voices die out, signaling the end of the mating season.

Frogs produce their calls in an interesting way. Closing their mouths, they push air from their lungs into their mouths, and then through special tubes into their vocal sacs. The sacs act as resonance chambers and give to each frog species its unique sound. Some frogs have only one vocal sac; others, such as leopard and pickerel frogs, have two. After the call is produced, the sac or sacs partially deflate, and more air is drawn into the lungs to produce the sound again. The calls of some species can be heard nearly a mile away.

Frogs sing usually from dusk until about midnight, and most sing their best when the temperature is between 55 and 70 degrees. More important to the volume of their chorus is the amount of rainfall during the mating season. Rainfall apparently stimulates their mating instinct. In fact, after a violent rainstorm late in the summer, male frogs that did not find a mate earlier may resume their singing.

The hind legs of frogs are much larger than the front legs. In toads, the difference in size is not so marked. Because of this, frogs are *leapers*—sometimes with surprising distance—while toads are *hoppers*, with shorter ranges. Both frogs and toads have moist skins, but toads generally are drier. Toads are covered with warts; most frogs have smooth wartless skins. Contrary to the myth, neither frogs nor toads produce warts in humans.

Both frogs and toads are carnivorous animals, with sticky tongues attached in the front of their mouths. The tongue folds outward with a quick darting motion to snap up insects and other small invertebrate animals.

A few of the nearly one hundred species of these amphibians are described here to help you recognize them. Others appear on pages 124 and 125.

ANDERSON'S TREE FROG is named after the South Carolina town where it was first discovered. It inhabits pine barrens, white cedar bogs, and sphagnum bogs of the eastern coastal plain. The frog measures a little under two inches and is green with a plum-colored sideband and yellow spots on its underside. Its unfroglike call is a nasal quack resembling the cry of the Virginia rail.

BULLFROG, seven to eight inches long and sometimes weighing as much as three pounds, is the largest frog in the United States. It is drab green on the back and yellowish white underneath. Its home is in the swamps and large ponds, where it feeds on snails, dragonfly nymphs, and May flies, sounding its deep bass *jug-a-rum more rum* on late summer evenings.

GREEN FROG, a solitary singer, is often mistaken for its larger neighbor in the swamps and ponds, the bullfrog, but the green frog may be distinguished by two parallel ridges running down its back from the eyes to the hind legs. The green frog, two to five inches long, is smaller than the bullfrog. It forages along streams and hibernates in natural springs, often wedging itself between stones.

GREAT PLAINS TOAD is a large, broad-bodied jumper, two to four inches in length. It has a gray-brown or greenish back, striped in the middle. This toad is found in the irrigating ditches, flood plains, and overflow bottom lands of the region that gives it its name. It breeds after rainfall, summoning the female with a shrill, piercing note.

OAK TOAD is a pigmy, three-quarters of an inch to an inch and a quarter in length. With its three to five pairs of brightly colored spots, it resembles a swatch of tapestry. The oak toad lives in the southern states, in such swampy areas as Okefenokee. It breeds in shallow cypress ponds, laying single bars of two to six or eight eggs on grass blades below the surface.

Endangered Wetland Wildlife

For many of the animals inhabiting the wetlands, the struggle for existence has been a losing one. Before the arrival of man it took species, in the normal course of evolution, thousands and millions of years to disappear. But in our time the destructive process has been speeded up considerably by man. Although man has learned how to change the world around him, he has not always used his knowledge wisely to conserve his natural heritage. For instance, in the relatively short history of the United States, some forty kinds of animals have vanished from the face of the earth, victims of man's effort to change his environment without considering all the consequences. The California grizzly, the San Gorgonio trout, and the handsome Labrador duck—all are gone forever. Today many other species are being threatened by the same fate.

The destruction of the marshes where the whooping cranes live, together with the wanton shooting of the birds by hunters, has reduced a once thriving population to only about forty individuals. Another bird, the everglade kite, is threatened by drainage and fire, which in combination destroy its sources of food. Some species are being harvested faster than they can reproduce; others, such as the snowy egret and the alligator, are being eliminated by greedy hunters who pursue the animals for their valuable plumes and skins.

When man is aroused and concerned, he can save life instead of destroy it. But to practice conservation wisely, he must learn which of his actions threaten other species. He must study how industrial pollution destroys a habitat, how drainage of a marsh imperils birds or fishes, how the introduction of one species will end the life of another.

Much has been done to change destructive courses and preserve our wildlife heritage. The establishment of national wildlife refuges is just one little step, but it is an important one. More such measures are urgently needed.

The animal and plant life shown on these pages are examples of those species whose existence is threatened.

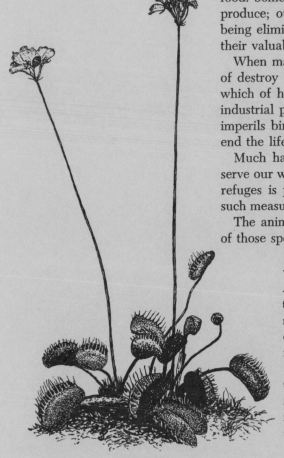

VENUS'S-FLYTRAP

Although this carnivorous plant with its glistening leaf is to some degree protected by law, it is nevertheless in mortal danger of extinction because clandestine violators continue to collect and sell the plant. The plant's future is further dimmed by its scarcity and its limited range (it grows only in the boggy soil of the coastal plain of the Carolinas). The trap is a two-lobed leaf that snaps shut when the plant's prey, usually small insects, touches its trigger hairs. The victim is eventually consumed by the digestive action of enzymes in the traps.

DUSKY SEASIDE SPARROW

There is some question whether this sparrow deserves to be classified as a distinct species. Its nearly black upper parts and the heavy black streaks on the underparts distinguish it from the Cape Sable sparrow. Living in fairly dry salt marshes, amid open mudflats, rushes, salt-hay grass, and glasswort, the dusky seaside sparrow is confined to the eastern coast of Florida in the vicinity of Merritt Island.

CAPE SABLE SPARROW

A single note followed by a buzzing trill signals the presence of this rare, isolated bird. Hurricanes, floods, and fires periodically sweep its flat, unprotected habitat—three square miles of brackish coastal prairies in Cape Sable, Florida—thereby depleting the bird's limited numbers.

FLORIDA SANDHILL CRANE

The existence of this gregarious and graceful water bird is threatened by hunters who take advantage of lenient local laws and also by the inability of this subspecies to reproduce in large enough numbers: its annual brood is two eggs. This long-necked, long-legged bird resembles in silhouette its fellow endangered species, the whooping crane. The sandhill is all gray except for its head, which is red and bare of plumage. Living mainly in the marshes of Alabama, southern Mississippi and Georgia south through the Florida Everglades, the Florida sandhill crane is a close relative of the sandhill cranes that live farther north and in parts of the west. Visitors to southern refuges delight in watching the whirling, hopping mating dance of this disappearing member of the North American wetlands.

ALLIGATOR

ALLIGATOR AND CROCODILE

These animals, now disappearing rapidly with the advance of civilization, are often mistaken for one another, probably as a result of their classification. Both belong to the order Crocodilia but to different families of the order. Actually, they can be distinguished very easily. The alligator (*above*) is darker and has a broader, more rounded snout; its teeth are completely covered when its mouth is closed. In contrast, the thinner crocodile (*below*) has its teeth exposed when its pointed snout is shut.

In the United States the two species live together only in Everglades National Park, Florida, although crocodiles are rarer in this region. Crocodiles live mainly in the marshes and mangrove swamps in the park, in the Florida Keys, and in Florida Bay. In all these places both species must struggle against extinction. Not only are they faced with the destruction of their swamp habitat by drainage and drought, but the alligators especially are in constant danger of hide poachers. Although the alligator is protected to some extent in Florida, Georgia, and Alabama, this protection is difficult to enforce; and the poachers continue to kill the young ones for purse latches, belt buckles, and letter-opener handles, among other "important" items. Many of the young alligators, perhaps most, will never reach breeding age.

There are precious few reptiles left in the world, and the alligators, crocodiles, and their relatives are the last of a group that has been around for about 200 million years. Man could destroy them in but a few years.

CROCODILE

One Reason Why Northern Plants and Animals Live in Bogs

For this experiment, you will need the following:

Two thermometers. The best kind is one that has the temperature scale etched into the glass. However, you may also use the kind of thermometer that is attached to a supporting back on which the scale is printed.

A handful of sphagnum moss. You can readily obtain this moss from a bog or, in the form of sphagnum peat moss, from a garden supply shop.

A board about a foot square.

Half a dozen thumbtacks.

An electric fan.

A small piece of screening.

Place the thermometers side by side on the board, about six inches apart. Note the readings on the temperature scales. They should be the same. Cover the bulb of one thermometer with a mound of moss, about an inch deep, but do not obscure the temperature scale. Secure the moss in place by covering it with the piece of screening, and fasten the screening to the board with the thumbtacks.

Sprinkle water on the moss until it is soaked. Wait five minutes, and then read the temperatures on the two scales. The moss-covered thermometer will have a lower temperature. Now wet the bulb of the uncovered thermometer with a drop or two of water. Position the electric fan so that it blows air on both thermometers and observe them carefully. The uncovered thermometer will go down rather quickly and then, as the bulb dries, return to very near its original temperature. The covered thermometer will begin to go down slowly, remaining at a lower temperature than the uncovered thermometer until the moss completely dries.

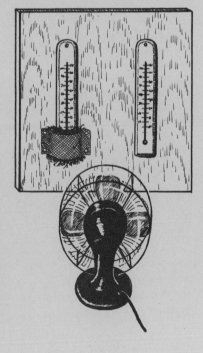

The uncovered thermometer went down rapidly because the water on the bulb evaporated quickly, and evaporation is a cooling process. The temperature of the covered thermometer went down slowly because it took a while for the cool surface of the moss to withdraw heat from the moss and water near the bulb. The covered thermometer remained cool because the moss was a reservoir of water which could rise to the surface of the moss by capillary action and then be evaporated to continue the cooling process for some time.

In a bog, during the hot summer, the carpet of wet sphagnum moss soaks up and evaporates huge quantities of water, thereby cooling the bog considerably. In contrast to the surrounding upland, the bog is a cool, moist habitat that favors northern plants and animals.

Glossary

Adaptation: An inherited characteristic that improves an organism's chances for survival in a particular *habitat*. Adaptations may involve the structure (form) or functioning (physiology) of an organism's body, as well as inherited behavioral patterns.

Alga (plural *algae*): The simplest of all green plant forms, having neither roots, stems, nor leaves. Algae range in size from microscopic single cells to branching forms one hundred feet or more in length. Larger marine forms are known as seaweeds.

Amphibians: The group of animals that includes frogs, toads, and salamanders. Amphibians have soft, moist skins and are characterized by life cycles in which the *larvae* usually live in water and breathe through gills, whereas the adults usually live on land and breathe through their skin and lungs but return to water to lay their eggs.

Amphipods: Scuds; a group of small *crustaceans* that have compressed bodies (flattened from side to side) and legs that can be used for both swimming and walking.

Arthropods: Animals with jointed legs and hard external skeletons. The group includes insects, *crustaceans*, spiders, and many similar animals.

Barbels: Fleshy threadlike sensory structures hanging like whiskers near the mouths of certain fish, such as catfish.

Beaver colony: A cooperative social unit composed of beaver parents and young. The parents limit the colony to about fourteen members.

Beaver lodge: Winter dwelling place of the beaver. Beavers build their lodges in summer out of branches, mud, and leaves.

Bivalve: Possessing two valves, or shells. Bivalve *mollusks* include oysters, clams, and similar animals.

Bog: A form of wetland, usually developing in a relatively deep lake with poor drainage. Bogs are characterized generally by extensive *peat* deposits, floating *sedge* or *sphagnum* mats, heath shrubs such as cranberry and leatherleaf, and often by the presence of *coniferous* trees such as black spruce and various cedars. *See also* Quaking bog.

Brackish: Used to describe waters that are mixtures of fresh and salt water. Coastal marshes and *estuaries* generally contain brackish, or moderately salty, water.

Byssal threads: Lines of silken, shiny material spun by the mussel from a gland in its foot in order to anchor its shell to a rock or bank.

Carapace: A hard shell-like covering on the upper side of an animal's body, such as can be found on a crab or turtle.

Carnivorous: Meat-eating; descriptive of organisms that feed on the flesh of animals.

Carrion: The dead body of an animal.

Chlorophyll: A group of *pigments* that produces the green color of plants; essential to *photosynthesis*.

Cilia (singular *cilium*): Microscopic filaments projecting from certain types of cells. Their rhythmic beating or vibration moves the cell itself or moves matter—for example, food—in the digestive system of the clam.

Community: All the plants and animals in a particular *habitat* that are bound together by *food chains* and other interrelations.

Competition: The struggle between individuals or groups of living things for such common necessities as food or living space.

Conifer (adjective *coniferous*): A plant that bears its seeds in cones. *See also* Evergreen.

Consumer: Any living thing that is unable to manufacture food from nonliving substances but depends instead on the energy stored in other living things. *See also* Decomposers; Food chain; Producers.

Crustaceans: The large class of animals that includes crabs, *amphipods*, isopods, and similar forms. Crustaceans typically live in water and are characterized by jointed legs, segmented bodies, and hard external skeletons.

Deciduous: Describing a plant that periodically loses all its leaves, usually in autumn. Only a few *conifers*, such as larch and cypress, are deciduous. *See also* Evergreen.

Decomposers: Living plants and animals, but chiefly *fungi* and bacteria, that live by extracting energy from the decaying tissues of dead plants and animals. In the process, they also release simple chemical compounds stored in the dead bodies and make them available once again for use by green plants. *See also* Mold.

Detritus: Minute particles of the decaying remains of dead plants and animals; an important source of food for many *marsh* animals.

Diatom: A single-celled *alga* encased in an intricately etched pair of silica shells formed by two halves that fit together like the lid on a box. Diatoms are important primary *producers* in the waters of the wetlands.

Dinoflagellates: A group of single-celled marine *algae* that possess characteristic plates covering their bodies and two *flagella* which control their movements.

Drainage pattern: The courses that water follows over a land mass as it flows to the oceans.

Ecology: The scientific study of the relationships of living things to one another and to their *environment*. A scientist who studies these relationships is an ecologist.

Ecosystem: A complex system of exchanges of materials and energy between living things and their physical *environment*. The system is repeated in cycles. It is also known as an ecological system.

Emergents: Plants, such as cattails and bulrushes, that root in the mud underwater and protrude above the surface.

Energy cycle: The process in which the energy of the sun is passed from one living organism to another. Green plants, the *producers*, capture solar energy. It is passed to the plant eaters, then to one or more of the meat eaters, and finally to the *decomposers*.

Environment: All the external conditions, such as soil, water, air, and organisms, surrounding a living thing.

Estuarine: Of or pertaining to an *estuary*.

Estuary: A tidal river; the portion of a river that is affected by the rise and fall of the tide and that contains a mixture of fresh and salt water.

Evergreen: A plant that does not lose all of its leaves at one time. Most North American *conifers* are evergreen. *See also* Conifer; Deciduous.

Evolution: The process of natural consecutive modification in the inherited makeup of living things; the process by which modern plants and animals have arisen through *adaptation* and natural selection from more generalized forms that lived in the past.

Eyes: Small areas of open water in *bogs*.

Flagellum (plural *flagella*): A whiplike structure used for locomotion by many single-celled organisms, such as *dinoflagellates.*

Floaters: Plants, such as water lilies, whose leaves float on the surface of the water but are connected by leafstalks or stems to roots are rooted floaters. Unrooted floaters, such as duckweed, are not attached, and the whole plant floats.

Flyways: Routes followed by migrating birds. In North America, administrators distinguish between the Atlantic, Mississippi, central, and Pacific flyways.

Food chain: The passage of energy and materials in the form of food from *producers* (green plants) through a succession of plant-eating and meat-eating *consumers.* Green plants, plant-eating insects, and an insect-eating fish would form a simple food chain. *See also* Food web.

Food web: A system of interlocking *food chains.* Since few animals rely on a single food source and since a given food source is rarely consumed exclusively by a single *species* of animal, the separate food chains in any natural *community* interlock and form a web.

Fossil: Any remains or traces of animals or plants that lived in the prehistoric past, whether bone, cast, track, imprint, *pollen,* or any other evidence of their existence.

Fungi (singular *fungus*): A group of plants lacking *chlorophyll,* roots, stems, and leaves. They must live off live matter as *parasites* or off dead matter as *saprophytes.* Mushrooms and water molds are fungi. *See also* Decomposers.

Glucose: A form of sugar produced by plants in *photosynthesis;* the main form in which carbohydrates are transported from cell to cell in animals and plants. Glucose is also known as dextrose and as corn or grape sugar.

Habitat: The immediate surroundings (living place) of a plant or animal.

Herbaceous: Referring to nonwoody plants whose aboveground parts wither away after each season's growth. Grasses, bulrushes, and cardinal flowers, among others, are herbaceous.

Hibernation: A prolonged winter dormant period during which an animal's body temperature becomes nearly that of the surrounding environment and its metabolic processes are extremely curtailed.

Incisors: Teeth adapted for cutting.

Invertebrate: An animal without a backbone. Insects and mussels are invertebrates. *See also* Vertebrate.

Kettle-hole lakes: Bodies of water formed in depressions left by melted glacial ice blocks. *Bogs* often form in these depressions.

Larva (plural *larvae*): An active immature stage in an animal's life history, during which its form differs from that of the adult. The caterpillar, for example, is the larva of a butterfly; the tadpole is the larval stage in the life history of a frog. *See also* Metamorphosis; Pupa.

Mammals: The class of animals that includes muskrats, rabbits, man, and many other warm-blooded creatures. They typically have a body covering of hair and give birth to living young, which are nursed on milk from the mother's breast.

Marsh: A treeless form of wetland, often developing in shallow ponds or depressions, river margins, tidal areas, and *estuaries.* Marshes may contain either salt or fresh water. Prominent among the vegetation of marshes are grasses and *sedges.*

Metabolism: The sum of the chemical activities taking place in the cells of a living thing.

Metamorphosis: A change in the form of a living thing as it matures, especially the transformation from a larva to an adult. *See also* Larva; Pupa.

Methane: Marsh gas; formed as a by-product of decomposing plant tissues.

Mold: A fuzzy fungal growth on dead plants and animals. Both water and terrestrial forms of molds help break down dead material. *See also* Decomposers.

Mollusks: A major group of animals with soft boneless bodies and, usually, shells. The group includes snails, clams, mussels, and oysters.

Muskeg: A mossy *bog* in the northern *coniferous* forest region.

Mycelia: The mass of threadlike filaments that composes the plant body of a *fungus*. The fungus attaches itself to a host by means of these filaments.

Nocturnal: Active at night.

Nymph: The immature, preadult form of the damsel fly and the dragonfly. Nymphs, after hatching, live and feed in water until they molt and become adults.

Panne: A shallow depression containing water left by receding tides. The water is usually too salty to support vegetation, but sometimes mats of blue-green *algae*, stunted grasses, or showy flowers form. Pannes and creeks are the principal physical features of salt-water *marshes*.

Parasite: A plant or animal that lives in or on another living thing (its host) and obtains part or all of its food from the host's body.

Peat: Partly decayed organic matter formed in boggy areas where high acidity and a lack of oxygen limits decomposition. *See also* Bog; Sphagnum.

Photic zone: Area penetrated by light.

Photosynthesis: The process by which green plants convert carbon dioxide and water into simple sugars. *Chlorophyll* and sunlight are essential to the series of complex chemical reactions involved.

Pigment: A chemical substance that imparts color to an object by reflecting or transmitting only certain light rays. *See also* Chlorophyll.

Pistil: Female part of a flower. During reproduction the pistil receives the *pollen* which contains the male germ cell. *See also* Stamen.

Plankton: The minute plants and animals that float or swim near the surface of a body of water. The enormous quantities of plant plankton (phytoplankton) and animal plankton (zooplankton) in water provide an important food source for many aquatic animals.

Pleistocene: Of or pertaining to the most recent epoch in the earth's history, roughly the past one million years. The period includes at least four major retreats and advances of continental glaciers.

Pollen: Tough, minute, grainlike bodies produced in the *stamens* of flowers or in staminate cones which contain the male germ cell. Blown or carried to a *pistil*, the pollen grain develops a tubelike outgrowth which penetrates to an ovule. The male germ cell moves through a tube into the ovule, where it unites with the female germ cell and fertilization occurs.

Predator: An animal that lives by capturing other animals for food.

Producers: Green plants, the basic link in any *food chain*. By means of *photosynthesis*, green plants manufacture the food on which all other living things ultimately depend. *See also* Consumer.

Productivity: The total number of living things in an *ecosystem*. Productivity is dependent on the interaction of the life and the *environment*.

Pupa (plural *pupae*): The inactive stage in an insect's life history when the *larva* is transforming into an adult. *See also* Metamorphosis.

Quadrifids: Small four-celled clusters on the inner walls of each sac of the animal-trapping bladderwort plant. The quadrifids absorb the water flowing into the submerged sacs, thus setting the trap.

Quaking bog: A bog whose floating mat lies unstably on water, so that pressure at one point produces shaking and trembling for a considerable distance around the point of impact.

Rootstock: A horizontal underground stem, sometimes confused with a root. Unlike roots, rootstocks have nodes, where buds and small leaves are usually attached.

Saprophyte: An organism, such as a mushroom, that obtains its food from nonliving matter. *See also* Decomposers; Fungus; Mold; Parasite.

Scavenger: An animal that eats the dead remains and wastes of other animals and plants.

Sedge: A kind of plant resembling the grasses. However, sedges usually have solid triangular stems in contrast to the round hollow stems of grasses. The floating mats of *bogs* are often composed of sedges.

Sessile: Permanently attached to a surface.

Shrub: A woody plant, usually less than twelve feet tall and having many stems rising from the ground.

Spat: Immature oysters newly settled upon an available surface.

Spatterdock: Yellow water lily; also known as cow lily.

Sperm: Male reproductive cell.

Sphagnum: A kind of light green, water-absorbent moss characteristic of *bogs*. *Peat* is comprised chiefly of the partly decayed remains of sphagnum moss plants.

Stamen: Male part of a flower. The stamen produces *pollen*, which contains the male germ cell. *See also* Pistil.

Standing crop: Biomass; the total weight of living matter in a *community* or of a particular kind of organism in a community. We may refer to the standing crop, or biomass, of a specific marsh or of ducks in that marsh.

Stomate: A microscopic opening in the surface of a leaf that allows gases to pass in and out.

Submergents: Plants that never reach the surface of the water. Pondweeds and certain forms of *algae*, such as *Chara*, are submergents.

Succession: The gradual replacement of one *community* by another, usually leading to a more or less stable community.

Swamp: A form of wetland characterized by moss and shrubs, or trees such as maples, gums, and cypresses. Swamps usually have better drainage than *bogs*. Sometimes they succeed *marshes* in shallow water basins, and they also may develop in sluggish streams and flood plains.

Territory: An area defended by an animal against others of the same species. It is used for breeding, feeding, or both.

Tuber: A fleshy or thickened underground stem, such as the edible portion of a potato plant.

Vertebrate: An animal with a backbone. The group includes fishes, *amphibians*, reptiles, birds, and *mammals*.

Water table: The upper level of the underground reservoir of water.

Bibliography

PLANT LIFE

DARWIN, CHARLES. *Insectivorous Plants*. D. Appleton, 1897.

FASSETT, N. L. *Manual of Aquatic Plants*. McGraw-Hill, 1940.

LLOYD, F. E. *Carnivorous Plants*. Chronica Botanica, 1942.

MATHEWS, F. SCHUYLER, and NORMAN TAYLOR (Editors). *Field Book of American Wild Flowers*. Putnam, 1955.

MOLDENKE, HAROLD N. *American Wildflowers*. Van Nostrand, 1949.

MUENSCHER, W. C. *Aquatic Plants of the United States*. Comstock, 1944.

POOL, RAYMOND J. *Flowers and Flowering Plants*. McGraw-Hill, 1957.

ROSE, FRANCIS. *The Observer's Book of Grasses, Sedges and Rushes*. Warne, 1942.

WHERRY, EDGAR T. *Wild Flower Guide*. Doubleday, 1948.

WODEHOUSE, R. P. *Pollen Grains*. Hafner, 1965.

ANIMAL LIFE

BENT, ARTHUR C. *Life Histories of North American Marsh Birds*. Smithsonian Institution, Bulletin 135, 1926.

BISHOP, SHERMAN C. *Handbook of Salamanders*. Comstock, 1947.

HYLANDER, CLARENCE J. *Fishes and Their Ways*. Macmillan, 1964.

KORTRIGHT, F. G. *Ducks, Geese and Swans of North America*. Stakpole Co., 1953.

MARTIN, A. C., and F. M. UHLER. *Food of Game Ducks in the United States and Canada*. Research Report 30, Fish and Wildlife Service, United States Department of the Interior, 1951.

MATTHIESSEN, PETER. *Wildlife in America*. Viking, 1965.

NEEDHAM, JAMES G., and MINTER WESTFALL. *Dragonflies of North America*. University of California Press, 1955.

RAND, A. L. *American Water and Game Birds*. Dutton, 1956.

STEBBINS, ROBERT C. *Amphibians and Reptiles of Western North America*. McGraw-Hill, 1954.

WRIGHT, ALBERT H. *Life Histories of the Frogs and Toads of the Okefinokee Swamp, Georgia*. Macmillan, 1931.

WRIGHT, ALBERT H., and ANNA A. WRIGHT. *Handbook of Frogs and Toads of the United States and Canada*. Comstock, 1933.

ZIM, HERBERT S., and CLARENCE COLTAM. *Insects*. Golden Press, 1963.

ZIM, HERBERT S., and HURST H. SHOEMAKER. *Fishes*. Golden Press, 1955.

ECOLOGY

BENTON, ALLEN H., and WILLIAM E. WERNER, JR. *Principles of Field Biology and Ecology*. McGraw-Hill, 1958.

BUCHSBAUM, RALPH, and MILDRED BUCHSBAUM. *Basic Ecology*. Boxwood, 1957.

CARPENTER, K. E. *Life in Inland Waters*. Macmillan, 1928.

DAUBENMIRE, R. F. *Plants and Environment*. Wiley, 1959.

ODUM, EUGENE P., and HOWARD T. ODUM. *Fundamentals of Ecology*. Saunders, 1959.

OOSTING, HENRY J. *The Study of Plant Communities*. Freeman, 1958.

REID, GEORGE K. *Ecology of Inland Waters and Estuaries*. Reinhold, 1961.

SHELFORD, VICTOR E. *The Ecology of North America*. University of Illinois Press, 1963.

TRIPPENSEE, REUBEN EDWIN. *Wildlife Management*. McGraw-Hill, 1953.

Water Fowl Management on Small Areas. Wildlife Management Institute, 1948.

FIELD GUIDES

BURT, WILLIAM H., and RICHARD P. GROSSENHEIDER. *A Field Guide to Mammals*. Houghton Mifflin, 1952.

COLLINS, HENRY HILL. *Complete Field Guide to American Wildlife: East, Central, North*. Harper and Row, 1959.

CONANT, ROGER. *A Field Guide to Reptiles and Amphibians of Eastern North America*. Houghton Mifflin, 1958.

LUTZ, FRANK E. *Field Book of Insects*. Putnam, 1935.

MORGAN, ANN HAVEN. *Field Book of Ponds and Streams*. Putnam, 1930.

MURIE, OLAUS. *A Field Guide to Animal Tracks*. Houghton Mifflin, 1954.

PALMER, RALPH S. *The Mammal Guide*. Doubleday, 1954.

PETERSON, ROGER TORY. *A Field Guide to the Birds*. Houghton Mifflin, 1947.

POUGH, RICHARD H. *Audubon Water Bird Guide*. Doubleday, 1956.

GENERAL READING

BUTCHER, DEVEREUX. *National Wildlife Refuges*. Houghton Mifflin, 1963.

CARRIGHAR, SALLY. *One Day at Teton Marsh*. Knopf, 1947.

CHAPMAN, VALENTINE J. *Salt Marshes and Salt Deserts of the World*. Wiley, 1960.

DARLING, LOUIS, and LOIS DARLING. *The Science of Life*. World Publishing, 1961.

ERRINGTON, PAUL L. *Of Men and Marshes*. Macmillan, 1957.

GABRIELSON, IRA N. *Wildlife Refuges*. Macmillan, 1943.

HOTCHKISS, NEIL. *Marsh Wealth*. Devin-Adair, 1964.

LINDUSKA, JOSEPH P. (Editor). *Waterfowl Tomorrow*. Bureau of Sport Fisheries and Wildlife, United States Department of the Interior, 1964.

MARTIN, A. C., H. S. ZIM, and A. L. NELSON. *American Wildlife and Plants*. McGraw-Hill, 1951.

ZIM, HERBERT S. *A Guide to Everglades National Park and the Nearby Florida Keys*. Golden Press, 1960.

Illustration Credits and Acknowledgments

COVER: Egret, Lester D. Line

ENDPAPERS: Sonja Bullaty

UNCAPTIONED PHOTOGRAPHS: 8–9: Bog development in northern Michigan, Lester D. Line 60–61: Black-crowned night heron, James A. Kern 104–105: Green frog, Lawrence Pringle 162–163: Cooters on log, James A. Kern

ALL OTHER ILLUSTRATIONS: 10–11: Vincent Cozzolino from G.A.I. 12: Graphic Arts International 13: William A. Niering 14–15: Simpson from Freelance Photographers Guild 15: Thase Daniel 16: Verna R. Johnston 17: Allan D. Cruickshank from National Audubon Society 18: Howard Sloane 18–19: James A. Kern 20: Graphic Arts International 21: Thomas A. Knepp 22–23: Charles Fracé 24: David C. Oschner; J. M. Conrader 25: William A. Niering 26: Danish Information Office 27: Patricia C. Henrichs 28: William H. Amos; Grant Haist 29: Monkmeyer Press Photos 30: Jack Dermid; Dr. William M. Harlow 31: Jack Dermid; Robert W. Mitchell 32: Oren S. Ryker; C. J. Stine 33: Robert W. Mitchell 34: Richard B. Fischer 35: Lester D. Line 36–37: W. J. Schoonmaker 38–41: Charles Fracé 42: Larry West from Full Moon Studio 43: Willis Peterson 44: Lynwood M. Chace 45: Glenn D. Chambers 46: David C. Oschner 47: Roche; Robert Strindberg 48: Leonard Lee Rue III 49: J. M. Conrader 50: John H. Gerard; David C. Oschner; Stephen Collins 51: Robert Strindberg; J. M. Conrader 52: Stephen Collins 53: Verna R. Johnston 54: Leonard Lee Rue III from National Audubon Society 54–55: J. J. Shamon from National Audubon Society 55: Ed Park; Allan D. Cruickshank from National Audubon Society 56–57: Charles Fracé 58: Thomas Martin from Rapho-Guillumette 62: Hans Zillessen from G.A.I. 63: John H. Gerard 64: Grant Haist 65: Hans Zillessen from G.A.I. 66: M. Woodbridge Williams 67–68: Patricia C. Henrichs 69: Betty Barford; Robert W. Carpenter 70: Lynwood M. Chace; Matthew Vinciguerra 71: Freelance Photographers Guild 73: Jack Dermid 74: Leonard Lee Rue III 75–76: Ed Cesar from Annan Photo Features 77: Ed Park 78: Glenn D. Chambers 79: Jack Dermid 80–81: David Morhardt from Full Moon Studio 82: Charles Fracé 83: Thomas R. Broker 84: Patricia C. Henrichs 85: Carl Struwe from Monkmeyer Press Photos 86: Walter Dawn 87–88: William H. Amos 89: Davis from Freelance Photographers Guild 90: Robert Strindberg; James A. Kern 91: Thase Daniel 92: Harris from National Audubon Society 93: Allan D. Cruickshank from National Audubon Society 94–95: Patricia C. Henrichs 96: Thomas Martin from Rapho-Guillumette 97: David C. Oschner 98: Thomas Martin from Rapho-Guillumette 99: Patricia C. Henrichs 100–101: Hans Zillessen from G.A.I. 102: Wilford L. Miller 106–107: Patricia C. Henrichs 108: John S. Flannery 109: Roche 110: William A. Niering 111: J. M. Conrader 112–113: Kelley Motherspaugh 114: Robert Strindberg 115: Ed Park 116: Edward S. Ross; Louis Darling 117: Louis Darling 118: Edward S. Ross 119: Patricia C. Henrichs 120: Larry West from Full Moon Studio 121: Stephen Collins 122: Leonard Lee Rue III; Jack Dermid 123: Hans Zillessen from G.A.I. 124: Lynwood M. Chace 125: Larry West from Full Moon Studio 126–127: Van Heussen from the United States Department of the Interior 128–129: Bruce Roberts from Rapho-Guillumette 130: Gibbs Milliken from Monkmeyer Press Photos 131: Glenn D. Chambers 132–133: Donald S. Heintzelman 134–135: Robert Strindberg 136: Donald S. Heintzelman 137: Michael Wotton 138: Jack Dermid 139: Hans Zillessen from G.A.I. 140–141: Patricia C. Henrichs 142–143: Allan D. Cruickshank from National Audubon Society 144: Hans Zillessen from G.A.I. 145–147: Leonard Lee Rue III 147: William J. Jahoda 148–149: Charles Fracé 150: Harry L. Beede 151: Allan Roberts 152: Ed Park 153: Karl H. Maslowski from National Audubon Society 154–155: Glenn D. Chambers 156: Charles Ott 157: Grant Haist 158–159: Monkmeyer Press Photos 160: Josef Muench 164–165: Stephen Collins 166: Graphic Arts International 167: Delaware State Development Department 168: Graphic Arts International 169: Hans Zillessen from G.A.I. 170: John and Jane Perry 171: Kesteloo from National Audubon Society 172: Fred M. Roberts 173: Berny Schoenfeld 174–175: Luther Goldman, National Park Service 175: Graphic Arts International 176–177: C. Evans, Fish and Wildlife Service 178–179: Freelance Photographers Guild 180: Allan D. Cruickshank from National Audubon Society 181: Hugh Halliday from National Audubon Society; Allan D. Cruickshank from National Audubon Society; Frank R. Martin, U.S. Fish and Wildlife Service 182: Michael Wotton 183: Walter Dawn 184: Edward S. Ross 185: C. Evans, U.S. Fish and Wildlife Service 186: Hans Zillessen from G.A.I. 187: George Komorowski from National Audubon Society 188–189: Robert Strindberg 190–191: United States Fish and Wildlife Service 192: New York Public Library 193: Bettmann Archives 194: Fur and Fashion Information Council 195: Gordon Smith from National Audubon Society 196–197: Michael Wotton 198: William A. Niering 201: Patricia C. Henrichs 203–209: Charles Fracé 210–211: Graphic Arts International 212–215: Charles Fracé 216–217: Patricia C. Henrichs 218–222: Charles Fracé

PHOTO EDITOR: ROBERT J. WOODWARD

ACKNOWLEDGMENTS: *The publisher wishes to thank Wayne W. Bryant, William Perry, and Orthello Wallis of the National Park Service, all of whom read the entire manuscript and offered valuable suggestions. The editorial assistance of Alice Grey, Department of Entomology, American Museum of Natural History, and of William H. Amos, ecologist, is also appreciated.*

228

Index

[Page numbers in boldface type indicate reference to illustrations.]